THE BEGINNER'S GUIDE TO RAISING GOATS

THE BEGINNER'S GUIDE TO

RAISING GOATS

HOW TO KEEP A HAPPY HERD

AMBER BRADSHAW

ROCKRIDGE
PRESS

Interior and Cover Designer: Elizabeth Zuhl
Art Producer: Sue Bischofberger
Editor: Arturo Conde
Production Manager: Jose Olivera
Production Editor: Sigi Nacson

Photography © Zoran Djekic/Stocksy, cover; evgenyatamanenko/iStock, pp. ii–iii; borchee/iStock, pp. vi–vii; Janet Horton/Alamy Stock Photo, p. x; RuralRoadsPhotography/Stockimo/Alamy Stock Photo, p. 18; Capuski/iStock, p. 26; ovbelov/iStock, p. 31; vgajic/iStock, p. 38; Tutye/iStock, p. 44; Angela DeCenzo/Alamy Stock Photo, p. 49; CasarsaGuru/iStock, p. 52; temmuzcan/iStock, p. 59; Jay Yuno/iStock, p. 64; Russotwins/Alamy Stock Photo, p. 79; mladenbalinovac/iStock, p. 83; Stephen Morris/Stocksy, p. 84; TFoxFoto/shutterstock, p. 98; Nirian/iStock, p. 108; Andre Babiak/Alamy Stock Photo, p. 121; Artem Shadrin/Alamy Stock Photo, p. 122; nikidavison/iStock, p. 134.

Illustration © ayutaka/iStock, pp. ix and 4–9; Bruce Rankin, p. 10; Netkoff/Creative Market, p. 104.

TO MY HUSBAND, TIMMY, AND
OUR THREE CHILDREN,
GAVIN, MORGAN, AND LINDEN

CONTENTS

INTRODUCTION: WHY YOU SHOULD RAISE GOATS

We began our goat journey almost a decade ago. My husband and I were having coffee one morning when he said, "You'll never believe this dream I had. I was standing on our porch with my coffee, and a goat walked up to me and started eating my robe!" We had never entertained the idea of owning goats. We lived by the beach, and my husband was a surfer, not a farmer. I had never tasted goat milk or goat meat—or even used goat fiber. Even though we were striving to live sustainably by raising bees and growing some of our own food on our tiny coastal homestead, livestock larger than a chicken wasn't a part of our five-year plan.

Lo and behold, a friend called me just days later and said she knew a woman who had two Nigerian Dwarf goats that were in desperate need of a home. Long story short, it's a decade later and we are still raising nine and counting. For me, it was love at first sight, but it took a while for my husband to come around. However, I always remind him how it all started with his dream.

While goats have been standard livestock for thousands of years, it's safe to say they have crossed over into the domestic realm. Nowadays, you'll see goats wearing everything from pj's to diapers. People are pushing them in grocery carts, and they even have their own YouTube channels.

I believe goats are so popular here in the United States because they are so stinking cute and they are more entertaining than any reality TV show. Although many have tried to domesticate goats into house pets like dogs and cats, goats are very much livestock and belong more on a bale of hay than on a couch.

Goats make ideal livestock for small and large homesteads, especially if you are looking to achieve a more self-sufficient and sustainable life. First, goats are truly multipurpose animals: They provide milk, cheese, butter, meat, fiber, leather, manure for your garden, and landscape services. Second, they can also be raised in a variety of climates and terrains—from the snowy mountains of Tennessee to the tropical beaches of South Carolina.

I wrote this book to help your goat herd thrive so you can reach your goal of living a more self-sufficient life and possibly make money from home. My husband and I may have jumped in with both feet and no experience, but it's a lot easier when you can rely on the experience and research of others—trust me! I've dealt with a lot of unnecessary heartache and trials. Fortunately for you, I've learned from my mistakes.

This book walks you through the basics of raising goats and sets realistic expectations about the time and commitment involved, so there are no surprises when you get your herd. It also shares ways you can monetize your goat herd—for fiber, meat, cheese, fertilizer, and more. It's important that any future goat owner know the pros and cons of raising goats—both for their happiness as well as their goats' happiness and health. Let your journey begin!

◄ Our first two goats were Nigerian Dwarfs named Wendy and Ziva.

chapter one

PICKING
YOUR GOATS

irst, some terminology: Female goats are called *does* or *nannies*. Intact males (males that have not been castrated) are called *bucks* or *billies*, while castrated males are called *wethers*. Young goats of both sexes are called *kids*. There are two different types of goats: mountain goats (which are wild goats) and domestic goats. This book covers only domestic goats, of which there are more than 200 breeds! Don't worry, we're going to stick to the most popular breeds here in the United States.

How do you choose from so many options? This chapter will explain how. It will also introduce you to domestic goat behavior and anatomy, as well as what you need to do to keep them happy and healthy. We'll also take a look at some practical tips for setting up a herd and purchasing your first goats.

WHY GOATS NEED COMPANY AND SHOULD NOT BE RAISED AS PETS

Goats are herd animals, which means that they evolved to live and feed together in a group. This is the environment in which a goat is most happy, and it's important to keep in mind when deciding on how many goats you should get. At a minimum, you should have two goats. I recommend three, so in the event that something happens to one of your goats, you still have the minimum recommendation of two.

While goats are cute, adorable, and very social, they aren't domestic pets, and I don't recommend you raise them that way. This doesn't mean you can't love your goats like pets—by all means, do. Even though goats can form a bond with humans, they should never be raised by themselves as a single goat. Humans aren't part of their herd. In fact, though some single goats may live happily with other hoofed animals (horses, sheep, etc.), there is no guarantee your goat will accept other species as part of its herd, so it's important to have two goats.

No matter how affectionate and attentive you are, studies have shown that goats who are raised without other goats can become lonely, depressed, and stressed. A stressed or depressed goat is more susceptible to sickness.

Our neighbor used to have a single goat as a pet. We saw her walking her goat on a leash down the streets in the neighborhood. It was adorable . . . until you heard the goat bleating all day and night whenever she left the house. It sounded like the poor thing was being tortured. Goats are like cookies: You need more than one.

WHICH BREED IS RIGHT FOR YOU?

There are three classifications of domestic goats: dairy goats, meat goats, and fiber goats. Some breeds are multipurpose, which means they can be good for both dairy and meat or meat and fiber. Each breed of goat has its own area where it shines. Your needs or goals determine the breed of goat you will want to raise, and this section highlights the best in class for each type of goat. We'll also cover other considerations, such as size, cost, and how and where to buy. In the end, you'll be equipped to make the best decision for your homestead or farm business.

TOP DAIRY GOAT BREEDS

Dairy goats are a wonderful alternative to cows for a smaller-scale dairy operation. They require less space and feed than their beef counterparts but still offer the benefits of milk, cheese, butter, and more. Outside the United States, goat milk is a preferred source for dairy consumption.

Depending on the breed and how often they are milked, does can stay in milk (produce milk) between 275 and 305 days a year, provided they have proper nutrition and health care. Although all healthy female goats can produce milk, only certain breeds are considered dairy breeds by the American Dairy Goat Association: Alpine, LaMancha, Nigerian Dwarf, Nubian, Oberhasli, Saanen, Sable, and Toggenburg.

These breeds are capable of producing enough milk for their own offspring and sharing with humans, but they vary in size and milk production. Here is a snapshot of each breed, but I recommend doing additional breed-specific research so you know what to look for when purchasing.

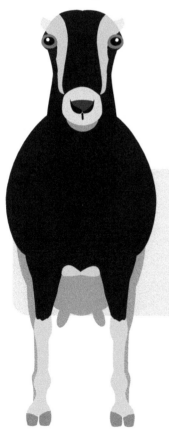

Alpine: These goats are second in dairy production only to Saanens, producing 2,620 pounds of milk per lactation. They are a medium-to-large goat weighing 135 to 170 pounds.

LaMancha: They are considered to be the fourth-best breed in milk production: 2,349 pounds per lactation cycle on average. Although their weight is comparable to that of Alpines (130 to 160 pounds), they are stockier. You can easily identify them by their tiny ears.

Nigerian Dwarf: These goats are the smallest of the recognized dairy goat breeds. With an average weight of 75 pounds and a height of no more than 23.5 inches, this breed is great for those with smaller plots of land or who need a smaller dairy animal.

Their milk production is at the bottom of the scale with an average of 813 pounds per lactation cycle. However, they are sought after for having some of the highest fat content in their milk.

Nubian: They are one of the larger goat breeds with a weight of 135 to 175 pounds. They also have a higher butterfat content than some of the other dairy breeds. Nubians can produce an average of 1,820 pounds of milk during their lactation cycle.

Oberhasli: Although they weigh between 100 and 150 pounds, Oberhasli are very strong for their size. They are heavy milkers, producing an average of 2,146 pounds of milk during their lactation cycle. Due to their brute strength, they may be tough to handle on the milk stand for beginners.

Saanen: These are the highest milk producers among dairy goats, producing an average of 2,765 pounds of milk per lactation cycle. They are one of the largest of the dairy breeds and weigh on average 132 to 198 pounds. They are light in color and sensitive to direct sun but ideal if you want a high-producing dairy goat.

Sable: These goats are a medium-large size, weighing an average of 145 pounds. Sables are heavy milkers, producing 2,570 pounds on average during their lactation cycle. They are basically Saanen goats that are not white.

Toggenburg: They are medium-large dairy goats weighing 120 to 200 pounds. Toggenburgs are good quality milk goats that produce 2,115 pounds of milk during their lactation cycle. Due to their thick coats, they thrive in cooler climates.

TOP MEAT GOAT BREEDS

Goat meat is heavily consumed throughout the world, and its popularity is steadily rising in the United States. It's leaner and lower in cholesterol than any other red meat. The United States imports over 50 percent of its goat meat, making this a viable business option.

Cabrito is the Spanish word for "baby goat" or "kid." In the United States, the term *cabrito* refers specifically to the kid meat (4 to 8 weeks old) that is used mostly for barbecued goat recipes. *Chevon* is another type of kid meat, from goats weighing 48 to 60 pounds that are 6 to 9 months old. Of these two types, cabrito is the most tender meat. Most goat meat is harvested when the goats are under a year old. Goat meat from goats older than a year tends to be tougher. Find more information about goat meat in chapter 9 (see page 99).

Boer: These goats grow fast. They are bred for their high growth rate and produce more meat in a shorter time than other breeds. Full-grown does weigh between 190 and 230 pounds, and bucks weigh between 200 and 340 pounds.

Kiko: These does weigh 100 to 150 pounds, and bucks reach 250 to 300 pounds. Kikos gain weight quicker than any other goat meat breed.

Myotonic: Also called fainting goats or Tennessee fainting goats, these goats are more parasite resistant than other breeds. Myotonic goats have a higher meat-to-bone ratio, though they are generally smaller than other meat breeds. The average weight is 60 to 175 pounds.

Savanna: This is a hardy breed. They were bred to survive in rough climates and poor forage. They are a well-muscled meat goat weighing 125 to 250 pounds.

Spanish: This breed was developed through natural selection in the wild. Due to this history, they tend to be smaller than other meat breeds. They are hardy and rugged and thrive on rough forage. Spanish goats weigh 50 to 200 pounds.

TOP FIBER GOAT BREEDS

Those who have worked with fine fibers will appreciate the importance of raising fiber goats. Cashmere and mohair fibers have been used in some of the finest fabrics for thousands of years. When considering raising fiber livestock, many think of sheep first. However, goats produce natural fibers that are durable, warmer than sheep wool, and beautiful.

The practice of raising goats for fiber has only been around in the United States a little more than 100 years, but its popularity is growing. The demand for these fine fibers has always exceeded their supply.

A micron (micrometer) is the measurement used to describe the diameter of wool fiber. The lower the micron value, the finer the fiber, so this measurement is important to determining its value. Mohair (from the Angora goat) has a micron value of 20 to 24, and cashmere (from various breeds) has a micron value of 14 to 19.

Angora: These goats produce mohair, and they are one of the most efficient fiber producers on earth. The United States is one of the two top mohair producers. Angora goats are calm and docile, but they aren't as hardy as the other fiber breeds. They weigh 100 to 225 pounds, and one Angora goat can yield 8 to 17 pounds of mohair a year.

Cashmere: These goats are not really a breed. Most goat breeds can produce cashmere; however, certain breeds and crossbreeds are considered "cashmere goats" due to their fine coats and the amount of cashmere they produce. This luxurious fiber outlasts wool, is a fraction of the weight, and is softer on the skin. Some of the popular cashmere goat breeds are Nigora, Pygora, and Kaghani.

Changthangi/Pashmina: These goats produce an ultra-fine cashmere and are usually found in cold, arid regions of the world, such as parts of India. Changthangi goats have been placed on the endangered species list by the National Bureau of Animal Genetic Resources. Once used for their meat, they are now praised for their fine down. Anything made with their exquisite fiber—which has a micron range of just 12 to 15—can command a high price. These goats are small, like Nigerian Dwarf goats, and average only 55 to 70 pounds.

UNDERSTANDING THE ANATOMY OF A GOAT

It's important to understand the anatomy of any livestock you raise. Animals can't tell us what they need, but when we understand their anatomy, we can often decipher what their bodies are communicating. It's also important to realize that livestock of any kind is a life commitment. If you decide to raise goats, be prepared to commit to 8 to 16 years to their care. Bucks have a shorter life expectancy and wethers have the longest. Here's what you need to know to help understand your goat's health and age.

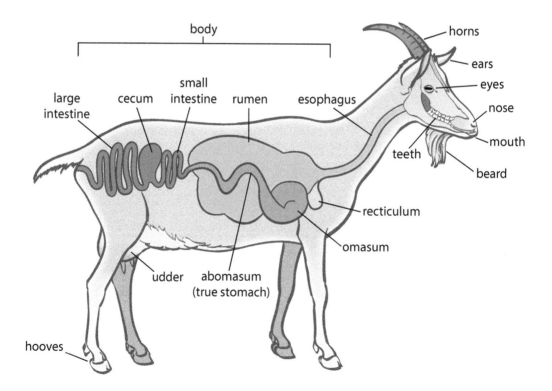

Beard: Both female and male adult goats have beards.

Body: Generally speaking, male goats are larger in build than females. Goats vary in appearance and may have several different colors in their coat.

Ears: The ears of a goat will be specific to the breed. For instance, LaManchas have almost no visible ears, while Nubians have very long ears.

Eyes: Goats have what some would call creepy slits or rectangles for pupils, instead of the round shape we are used to. Animals with side-slanted eyes, like goats, are what's called "grazing prey." The slanted pupils help extend their peripheral vision and detect predators.

Hooves: Goats' hooves are cloven, meaning they're split into two. They also have dewclaw hooves higher up on the back of their legs, which assist in traction.

Horns: Not all goats have horns; some are born without them. A goat that is born without horns is called *polled*.

Nose: Much like ears, a goat's nose shape is specific to its breed.

Udder: A goat's mammary gland is often referred to as an udder. Goats lactate like any other mammal; therefore, they produce consumable milk. Attached to the udder are teats, which is how the milk is drawn from it.

DIGESTIVE SYSTEM

Goats are herbivorous ruminants, which means they eat vegetation (grasses, fruits, leaves, vegetables, roots, hay, limbs, and bulbs) and have four-chambered stomachs. Each of these chambers helps to break down and digest their food so they can receive the most nutritional value from what they are eating.

Mouth: Goats lack upper teeth, so they primarily rely on their lower teeth, tongue, and lips to take in food.

Esophagus: Just like our esophagus, a goat's esophagus transfers food from the mouth to the stomach, or in a goat's case, the first part of its four-chambered stomach. It also transports gases and cud (food that returns to the mouth to be chewed again).

Reticulum and rumen: These are the first two chambers of the stomach, and together they provide a place for bacteria to break down food, especially fibrous food like hay, grass, leaves, and the like. Some energy absorption happens here, too.

Omasum and abomasum: Undigested food from the reticulum and rumen travels to the omasum. Commonly referred to as the "true stomach," this compartment functions similarly to human stomachs, with acid helping to break down food. The abomasum is the fourth and second-largest compartment of a goat's stomach. It receives food from the omasum and then passes it to the small intestine.

Small intestine: A goat's small intestine absorbs nutrients, similarly to our own.

Cecum: Also known as the blind gut, the cecum connects the small and large intestines. Microorganisms help digest food here, too.

Large intestine: Undigested food moves on to the large intestine, where liquid is absorbed, a similar process to what happens in our own large intestine.

HOOVES

The hoof is an important part of a goat's anatomy, and hooves will require maintenance. A goat's hooves grow indefinitely, just like our fingernails. The rate of growth varies from goat to goat and is influenced by breed, diet, exercise, and living conditions. As hooves grow, it is possible for them to curl and harbor bacteria, causing infection. In general, it's time to trim when hooves begin to curl either outward or inward. Depending on their severity, hoof problems may result in a decline in performance and health.

TEETH

Unless you've seen them smile, you would hardly guess goats don't have any upper teeth. All their teeth are located on the bottom. The top jaw is a hard dental plate that helps them digest their forage. Like humans, goats have two sets of teeth. Baby teeth, called milk teeth, come in from birth to 12 months, and permanent teeth come in from 1 to 4 years old, or until all 8 incisors have come in.

HORNS

The majority of goats have two horns. However, some are polled—born without horns. Horns are made up of both keratin and bone. They are narrow and have large blood vessels running up them. Goats use their horns to regulate temperature, protect themselves from predators, and assert their dominance in a herd.

WHERE TO GET YOUR KIDS

Once you've narrowed down the breed of goat you want to raise, you need to find a place to buy your goats. This part is very important, because you want to get good stock. There are a lot of wonderful breeders out there, but there is an equal number of bad breeders and unethical goat owners. This section will help you find the former rather than the latter.

Goat people know goat people. The best people to ask about where to find a breeder are other owners of the breed you want. If you don't know anyone who raises goats, contact your local farm extension agency, which will be familiar with your area.

Although you can find livestock for sale on social media and at livestock auctions, proceed with caution unless you have a reputable referral. Many people use these outlets to dump unwanted, or even sick, livestock.

LOCAL FEED STORES

Anyone who owns livestock has to visit a feed store sooner or later. Many feed stores have a bulletin board for their customers to use. Breeders will often post a flyer if they have livestock or other farm goods for sale. You can also ask the sales clerks at the feed stores if they know of any reputable goat breeders in the area. My husband and I have sold livestock through local feed stores.

If your feed store doesn't have any leads, ask them for a recommendation of an equine vet or one who works with goats. A veterinarian will know who raises goats in the area and may even be able to tell you what type of breeder they are and give you some recommendations.

REGISTRIES AND GOAT CLUBS

In addition to knowing what kind of breed you'd like to buy, you also need to decide whether you want a registered or unregistered goat. A *registered goat* is one whose bloodline has been certified as pure by a recognized professional goat organization, such as the American Dairy Goat Association (ADGA), the American Goat Federation (AGF), or the American Goat Society (AGS). There are also breed-specific organizations, such as the Nigerian Dwarf Goat Association (NDGA). These organizations will have lists of participating members, so you can find a breeder who sells registered goats near you.

Owning a registered goat means you can see its lineage and ancestral history, enter it into shows and competitions, and participate in milk production performance programs. Registered goats can also command a higher selling price than those that are unregistered.

An *unregistered goat* may still have a pure bloodline, but it won't be verified and you may not be able to participate in shows and other goat programs. You will not be able to register a goat (or its offspring) in the future if you don't do it at the beginning. There are exceptions to this, but it's a long process that can be expensive. If you aren't interested in those things, an unregistered goat may be a good option; you can typically buy one for half the price of its registered counterpart. It's also a quicker transaction, because there's no need to locate a registered breeder.

It's important to note that there is no way to prove the bloodline of an unregistered goat. I once bought a "pure" Nigerian Dwarf goat from a "reputable" breeder without papers to save a couple of bucks. To my dismay, our goat was a mixed breed and there wasn't a darn thing I could do about it. Our mixed breed became a pet because I wanted a purebred, and from that point on, we only purchased registered goats with papers in hand.

BEFORE YOU BUY

They say that an ounce of prevention is worth thousands in goat vet bills, or something like that. If you take away anything from this book, let it be this: Make sure your goats are healthy *before* you buy them or take them home. Trust me on this. A good breeder will never be offended if you request a goat vet health check (at your expense) before you buy. If they are, then they have something to hide and you need to find a new breeder. A $75 wellness check is a lot cheaper than a $1,500 sick goat vet visit.

Find a veterinarian who specializes in the goat breed you're buying. You may not be able to use the same vet who treats your dog, Fido, nor should you want to. Goats are livestock and will need a livestock vet to treat and diagnose them in times of need. Next, plan a wellness visit when the goats are available for sale. If they get a clean bill of health, it's time for you to bring your goats home.

HOW TO EVALUATE YOUR GOAT

No matter what, you should follow my advice about seeking a professional wellness check before you buy. I did not heed my own advice and ended up spending thousands on sick goats and contaminating my virgin land with parasites. However, this isn't a perfect world and not every situation is perfect. There may come a time when you need to evaluate goats on your own. The information in this section will show you how to tell if a goat is sick or healthy and alert you to red flags before you call the vet.

The term *clean herd* refers to a herd that has been tested for sickness and disease. Most breeders will test their herd two to four times a year. A breeder can send off samples from their herd to a lab for their test results or have a vet come and perform these tests. Common health tests for goats include fecal tests for parasites, CAE (caprine arthritis and encephalitis), Johne's disease, TB (tuberculosis), CL (caseous lymphadenitis), and brucellosis. (For more on common illnesses, see page 67.)

Always ask your breeder if their herd has been tested recently (within six months of the proposed purchase date) and for a copy of the test results.

GET A GOAT WELLNESS CONTRACT!

A goat wellness contract is not a standard legal document but rather a written statement given by a breeder ensuring a goat's good health. It allows you to bring back the goat within a certain number of days if that goat becomes sick or dies, in exchange for another goat or a refund.

It would be nice if all goat breeders were people of their word and an old-fashioned handshake were all you needed to seal the deal. Unfortunately, that is not the case. Reputable breeders *should* offer you a health contract, as well as the registration papers (if you're buying registered goats), at the time of purchase. I strongly suggest that you do not bring a goat home without both the wellness contract and the registration papers. You're just going to have to trust me on this one.

GOOD HEALTH CHECKLIST

Here are the basics of what to look for when evaluating the general health of a goat. Again, it's always best to go off the guidance of a vet trained to work with goats. If the goats you want to buy are showing any sign of sickness or look unhealthy, walk away. Better yet, run away. A sick goat/herd can quickly become a money pit.

- ○ **Eyes:** Should be clear, not runny, red, or with any sign of infection.
- ○ **Nose:** Should be soft and moist with no discharge.
- ○ **Ears:** Should be clean and free of buildup or mites.
- ○ **Mouth:** Should be pink, not white, red, or with foul odor.
- ○ **Hair:** Should be soft and shiny with no missing patches.
- ○ **Legs:** Should be straight and have an even distribution of weight when the goat is walking.
- ○ **Hooves:** Should be trimmed and clean; bottoms should be clear of any sign of infection.
- ○ **Rear end:** Should be clean and dry with no signs of loose or runny stool.
- ○ **Stool:** Should be firm round pellets and not leave any visible sign on the rectum.
- ○ **Personality:** Should be energetic and personable, not lethargic or spooked.

chapter two

PREPARING YOUR HOME

You did your research, picked your goat breed, and found a veterinarian. Now it's time to get the homestead ready for your new goats. This chapter will cover everything you need to get prepared for your new goat adventure. A happy, healthy goat herd requires a bit of effort when it comes to housing, including fencing, shelter, and protection from wildlife.

HOW MANY KIDS CAN YOU RAISE?

Because goats should not be raised alone, they will need the company of other goats, or at least other hooved animals. How many you can raise will depend on the size of your goats and how much space you have available to raise them.

Kids weigh about 4 to 12 pounds and full-grown goats between 50 and 250 pounds. If you are starting your herd with baby goats, you won't need to provide a large amount of living space immediately, but you do need to expand their area as they grow.

As a rule of thumb, each goat requires a minimum of 20 square feet of indoor living space and 250 square feet of yard space. So that means if you have three goats, their shared indoor space must be at least 60 square feet and their outdoor space must be at least 750 square feet.

But not all types of goats can live together. Does and bucks should not be raised together beyond 16 weeks of age. Good goat combinations are does with does and wethers, bucks with bucks, or bucks with wethers. If you plan on breeding goats, then you will need a separate area for the does and bucks. This is to prevent unwanted mating and pregnancy, and to maintain milk quality (when goats are in heat or rut, it can affect the taste of their milk).

Goats prefer to browse and forage on fresh vegetation; they won't graze on grass unless they don't have any other food available. They don't do well in mud or wet soil, as this can cause hoof issues. Protection from the wind and rain is also a must. Their ideal home has access to fresh trees, shrubs, brush, weeds, and dry land; areas for climbing and playing; a fenced-in yard; a feed area that is protected from the rain; and a shelter with shelves to climb and lie on. Bonus spaces include a birthing and milking area.

GOAT-PROOFING YOUR HOME WITH FENCING

Goats need to be protected from the things that may harm them on your property—and your property may need to be protected from them. That's why, just like you'd baby-proof your home, you need to goat-proof your property. Whatever goats have access to could possibly become a chew-thing. Our goats have eaten my husband's favorite flip-flops, the cedar siding on our home, insulation, and more. Not only can they chew on things they aren't supposed to, but there is also a list of forage that can make them sick (see "Protecting Your Herd" opposite).

TYPES OF FENCE

There are many different types of fencing you can use for your goats, such as wood, electric, and wire. I've even seen people make goat fences with wood pallets. Our first goat fencing was six feet tall, made up of wood fence panels, which were the perfect height. Now we have welded wire, and I enjoy having the ability to see the goats at all times through the wire.

A good fence should do two important things: It should keep your goats in and predators out. It can also be crucial when you need to separate bucks from does, for creating birthing areas, or for separating a sick goat from the herd.

Goat fencing should be at least four feet high and have openings no bigger than four inches to reduce the chance that their head and horns will get stuck—which can be a problem. Four feet is an ideal height unless you have a large goat breed or have goats that are high-strung. If you have a larger goat breed, aim for a six-foot-tall fence.

While electric fencing is good for keeping your goats in and predators out, it should not be used by itself. Instead, use it in conjunction with other fencing. Wood fencing is the most expensive of the available options, but it is aesthetically pleasing. If you have your own mill and lumber, you can save a lot of money by building your own fence.

OTHER FENCING NEEDS

You may also need to use fencing to protect your property from your goats. For instance, our goats love jumping on our air-conditioner unit, which was damaging the coils and fan, so we built a fence around it to keep our goats out.

Goats love to browse on all things wood—trees, bushes, bark, and houses. They rub their horns and backs on wood and make an afternoon snack of it. If you don't want these things destroyed, you'll need to put up a barrier between them and your goats.

We still have a tree with memories of our goats permanently etched on its trunk. Goats do more extensive damage to tree bark if they don't have enough forage or roughage. However, even in ideal settings, they still may take a fancy to your favorite tree. Utilize wire fencing to help protect the trunks of your trees. Do this by wrapping the trunk of the tree with wire high enough that your goats can't reach it.

PROTECTING YOUR HERD

The first step toward protecting your herd is learning about common predators in your area. It is crucial to keep your herd safe by locking them up at night in order to prevent predators from getting easy access. Predators vary by region. Some of the predators you may need to be on the watch for are:

- Bears
- Bobcats
- Coyotes
- Dogs
- Eagles
- Foxes
- Hawks
- Mountain lions
- Wild hogs
- Wolves

For instance, in my area, we have coyotes, bears, foxes, and bobcats. In order to ensure the safety of your goats, you may also consider adopting a livestock guardian animal, which are either trained to protect or have natural instincts that protect a herd.

We adopted two livestock guardian dogs into our family. Our dogs are a breed called Great Pyrenees, and they range from 80 to 120 pounds. In addition to having livestock guardian dogs, we keep our goats in an enclosed area that we are able to lock up and secure at night.

Livestock guardian animals are not pets; they are working animals. They have a job to do and will do it well with the proper training. Livestock guardian animals are worth their weight in gold, and I strongly suggest that anyone raising livestock invest in one or more. Other livestock guardian animals include donkeys, llamas, and guineas. Some even say geese can be livestock guardians!

On top of predators, another threat to your furry friends is toxic plants, shrubs, and trees. Although I suggest that you do a thorough inspection of your property for toxic plants, I don't think it's necessary to eliminate every single plant goats are sensitive to (unless it is a very toxic plant). We have a lot of mountain laurel, for example, and it would be impossible for us to rid all our land of it. I make sure our goats always have enough safe things to eat, and I keep an eye on them. Here's a short list of plants that are toxic to goats:

- African rue
- Azalea
- Boxwood
- Bracken fern
- Burning bush berries
- Chinaberries
- Chokecherry, wilting especially, or its leaves in abundance
- Datura
- Dog fennel
- Dog hobble
- Euonymus bush berries
- False tansy
- Flixweed
- Fuchsia
- Holly trees/bushes
- Honeysuckle
- *Ilysanthes floribunda*
- Japanese yew
- Lilac
- Lily of the valley
- Milkweed
- Monkshood
- Mountain laurel
- Oleander
- *Pieris japonica* (extremely toxic)
- Red maple
- Rhododendron
- Rhubarb leaves
- Virginia creeper
- Wild cherry
- Yew

CREATING AN ACTION PLAN FOR YOUR GOATS

Most people who start raising goats do so in order to live a more self-sufficient lifestyle, and unless you're able to monetize your herd immediately, you will have out-of-pocket expenses.

We were fortunate enough to already have a fenced-in yard when we first started raising goats, and we were able to repurpose a shed for their manger, so we saved some money on our initial start-up costs. You may be able to repurpose existing structures yourself to save money. I've seen people use pallets and tin, old playhouses, and even an old school bus as goat houses.

If you are investing in goats in hopes of making a profitable business or even a side business, be prepared to invest at least one year of your time before you'll see a return. Your first year's worth of start-up costs could be $2,000 to $12,000 and up, depending on where you live and how much work you are able to do on your own. Here is an example expense list. The prices listed below are per goat unless otherwise noted.

- Goat fee: $40 to $1,000 (the high end for a specific, show-quality goat)

- Vet wellness check: $50 to $150

- Bedding (hay or wood shavings): $10 per month per shelter (see page 29) ($120 per year)

- Minerals: $10 to $20 per month ($120 to $240 per year)

- Hay feed: $3 to $15 per 50-pound bale (feeding 2 to 4 pounds per day) ($44 to $440 per year)

- Food: $17 per 50-pound bag per month (feeding half a cup per day) ($204 per year)

- First aid kit: $60

- Housing: $0 to $10,000+

- Fencing: $500 to $5,000

Your goats will need time to get established on their new land, and you will need time to get used to them. Depending on whether you purchased kids or adults, they may need to mature to breeding age. Then when the kids are born, you will need to wait until they are weaned to either sell the kids or separate them so you can milk the mama. For more specific information on this time-line, see chapters 7 (page 65) and 8 (page 85).

chapter three

BUILDING SHELTER AND FENCING

Your goats need a safe place to call home, a place that protects them from the elements and predators. A goat shelter—or what we like to call a goat manger—can be as elaborate or as simple as you want. You may need to set up a temporary shelter when you first get them, but long-term housing plans should begin before you purchase your herd. This chapter covers everything you will need for your shelter, including step-by-step instructions for building your own and advice on bedding, ramps and shelving, birthing areas, and more.

ESSENTIAL COMPONENTS OF A SHELTER

In addition to keeping your goats safe from predators (see page 23), your goat shelter also needs to keep your goats dry and out of the elements. At the very least, this means goats need a roof and three walls. Recall, too, that no matter what your shelter looks like, each goat in your herd should have at least 20 square feet of living space inside the shelter. Let's take a closer look at the essential components.

Roof: Every goat shelter needs a roof that is free of leaks.

Walls: The walls should be sturdy enough to keep out wind and rain.

Flooring: The best flooring for a manger is dirt or gravel. Concrete is too cold and wood absorbs odors. Dirt and gravel allow the manure to disperse and decompose. On top of this, you need a couple of inches of either hay, wood shavings, or straw to help provide warmth and absorb smells (see page 30).

Room and board: If you plan to breed goats or keep both does and bucks, you need separate living quarters. The dividers don't need to be fancy, but they must be goat-proof, meaning they separate your goats so they don't have access to each other. You can use wood pallets, wire fencing, wood planks, or tin to divide stalls.

Birthing area: If you plan to breed goats, you'll need a birthing area that is separate from the main living quarters, away from the rest of the herd. This area provides the mama privacy, safety, and time to bond with her kids. This area should be a minimum of four feet by six feet; well insulated with hay, straw, or wood chips; and free from drafts and the elements. You will also need room for feed and water buckets.

SLEEPING AND BEDDING

Bedding helps keep your goats dry, absorbs ammonia from their manure, and provides warmth. You want something that is absorbent, soft, and cost-effective. Some common bedding materials include:

➼ **Hay or straw.** This is my preference. It's not as absorbent as the other two, but I always have it on hand for feeding the goats and pigs.

- ➡ **Pine shavings.** Make sure it's pine and not cedar, since cedar can cause issues with goats.

- ➡ **Wood pellets.** These are the same pellets used for horse bedding or wood pellet stoves.

Add a couple of inches of bedding to the entire floor of your goat shelter and on the top of any shelf they lie on.

KEEPING THE SHELTER CLEAN

How many goats you own and the season will determine how often you need to clean out the shelter. We have nine Nigerian Dwarf goats, and we clean out their shelter biweekly in the summer to help keep the fly population under control. We use the deep-bedding method in the winter, which means I just keep adding fresh bedding to their existing bedding without removing the old bedding. The decomposition of the old bedding gives off heat, which helps keep the goats warm in the winter months.

Your goat shelter should smell "goaty" or musty. It should never smell like ammonia or burn your eyes. If it does smell like ammonia, it is time to clean your goat shelter and replace the old bedding with new bedding.

HOW TO BUILD A TEMPORARY SHELTER

Skill Level: Beginner | **Estimated Material Cost:** $100 or less | **Time:** 2 hours

This plan is for a shelter that should only be used for temporary housing, until the permanent shelter (see page 34) can be built. You may also need a temporary shelter when moving goats from one area to another on your land or when separating goats for one reason or another. This shelter should have a dirt floor.

SUPPLIES AND TOOLS

- 2 (4' × 16') welded wire cattle panels
- Heavy-duty zip ties
- 5 (4-foot) T-post stakes
- 1 (16' × 20') heavy-duty tarp
- Hammer

Troubleshooting:

Goats are curious characters, and they like to eat things they shouldn't, including the tarp for their housing. Try to make sure to secure all loose ends and not have any overhang that are not secured or they will eat it.

1. Lay both cattle panels flat on the ground side by side, so that they cover an 8 × 16–foot area. Then push one panel over the other, so that it overlaps by 4 inches (one square of the panel). Use the zip ties to secure the panels together every three squares or so.

2. Using two of the T-posts, drive down each post into the ground about one foot deep where you want one side of your shelter. Make sure the posts are firm and solid in the ground and on either end of the 8-foot section.

3. Having someone help you, place the shorter end of the cattle panels up against the T-posts that are secured in the ground.

4. Hold the opposite end of the cattle panels and slowly walk the end toward the side that is pressing against the T-post to form an arch with the panels.

5. After the arch is made (arch should be about nine feet wide), one person continues to hold the arch in place while the helper drives the two remaining T-posts into the ground on the outside of the panel to secure the loose side.

6. Cover the arch with the tarp and secure it to the panels with zip ties. Make sure to completely cover the back and sides of the shelter, leaving the front open.

7. Add straw to the floor and place a wood shelf or folding plastic table inside for your goats to lie on.

HOW TO BUILD A PERMANENT SHELTER

Skill Level: Beginner | **Estimated Material Cost:** $500 or less | **Time:** 2 days

This shelter takes into account the long-term needs of your goats and will accommodate a bigger herd in the future. You'll start by framing 3 sides. Next, you'll work on the roof. The 2" × 4" × 10' boards are your roof rafters. The roof will be elevated in the front to help with runoff, and you will have a 1-foot overhang on the front and back. This shelter will be wide enough that you can divide it into separate living quarters if needed. If you build this shelter using wood screws instead of nails, it is possible to disassemble the structure and move it if you need to. This shelter should have a dirt floor.

SUPPLIES AND TOOLS

FOR THE FRAMING

- Disposable paintbrush
- Tar
- 3 (4" × 4" × 8') wood posts
- 3 (4" × 4" × 10') wood posts
- Post hole digger
- Level
- Measuring tape
- 2 (2" × 6" × 16') boards
- 1 box (3½-inch) decking screws
- Electric drill
- 4 (6' × 8') wood fence panels

FOR THE ROOF

- 10 (2" × 4" × 10') boards
- 6 (1" × 4" × 16') boards
- 8 (26" × 10') tin roofing sections
- 1 bag tin screws

TO MAKE THE FRAME

1. Measure and mark off an area for your shelter. Each side will be 8 feet long and the front and back will each be 16 feet wide. Choose a place that is flat and level with the opening facing south, if possible.

2. Using the disposable paintbrush, paint the bottom 2 feet of each 8-foot and 10-foot 4' × 4' post with the tar, then discard the brush. The tar protects the wood post from rotting in the ground.

3. Pretend you are facing your goat shelter. You are going to start your build at the back-left corner.

4. Using the post hole digger, dig a hole 2 feet deep. Place one of the 8-foot posts, tar-side down, in the hole.

5. Use the level to check for plumb on the side and front of the post. Once you are plumb, backfill the dirt and pack it firmly around the post. Check for plumb again.

6. From the outer corner of the post, measure over to the right 8 feet for the second hole.

Repeat steps 4 and 5, placing the second post at 8 feet center (the center of the second post should be eight feet from the outer corner of the first).

7. From the center of the second post, measure out 8 feet to the right. Repeat steps 4 and 5, using the last 8-foot post.

8. Pull the measurement from the outer edge of the third post to the outer edge of the first post; this total measurement should be 16 feet.

9. Measure from the outer edge of the first post forward 8 feet. Repeat steps 4 and 5 using a 10-foot post. This will become the corner post for the front of your shelter.

10. Measure from the outer left corner 8 feet. Repeat steps 4 and 5 with the second 10-foot post, placing it at 8 feet center.

11. From the center of the second post, measure out 8 feet to the right. Repeat steps 4 and 5 using the last 10-foot post.

12. You now have six posts in the ground. The three back posts are 6 feet high and the three front posts are 8 feet high.

13. Position one 2" × 6" against the back side of the 4" × 4" posts, flush with the top of the posts, and secure it with the decking screws; this is the shelter's back cross support.

14. Secure the remaining 2" × 6" board at the top of the front 4" × 4" posts.

15. The top of each 2" × 6" board should be flush with the top of your 4" × 4" posts.

16. Install the fence sections to the back and sides of the 4" × 4" posts using decking screws. You now have all three walls installed on your shelter.

CONTINUES

Troubleshooting:

It's important that the roof is watertight. The tin screws have a rubber gasket on them to help prevent water leaks. You can add some silicone caulk to the outside of the screws for an extra level of water protection on the roof.

TO MAKE THE ROOF

17. Starting with the front left, lay the first 2" × 4" on top of the two 16-foot cross-support boards, leaving a 1-foot overhang on both sides of the shelter (front and back). Secure it with decking screws. Make sure to line up the outer edge of your 2" × 4" with the outer edge of the 16-foot cross support.

18. From the outside left of the first rafter, measure over 2 feet and install the second 2" × 4" as you did in step 1. Repeat this process with the remaining 8 (2" × 4") boards, until you have 10 rafters.

19. Next, install the tin supports, which lie lengthwise across the rafters. Install the first 1" × 4" board flush with the front edge of the front rafter and secure with decking screws.

20. From the front edge of this first tin support, measure backward 2 feet and install the second 1" × 4" board and secure it with decking screws. Repeat this process with the remaining 4 (1" × 4") boards, until you have 6 tin supports.

21. Secure the tin roofing sections to the supports using tin screws.

HOW TO CLEAN YOUR GOAT SHELTER

Skill Level: Beginner | **Estimated Material Cost:** $10 per month per 8' x 16' shelter | **Time:** 30 minutes

SUPPLIES AND TOOLS

- Dust mask or respirator
- Gloves
- Shovel
- Large trash can or wheelbarrow
- Fresh bedding (pine shavings)

Troubleshooting:
If the smell is a little too goaty to you, add some lime or diatomaceous earth to the floor after cleaning and before you add the bedding. Just make sure to wear a dust mask and to open the doors to ventilate the shelter. I like to add mint sprigs to bedding as well. The mint adds a fresh scent and repels spiders.

1. Put on your mask and gloves, and make sure all the animals are out of the goat shelter.
2. Shovel out the old bedding, placing it into the trash can or wheelbarrow for composting.
3. Remove any spiderwebs and inspect for rodents.
4. Once you've removed all the used bedding, let the shelter air out for about 1 hour.
5. Add a couple of inches of new bedding. Compost the manure and old bedding.
6. Rinse off the shovel and spray with a disinfectant such as bleach.

chapter four

FEEDING YOUR GOATS

oats have long had a reputation for eating anything. Need your grass cut? Get a goat. Hedges trimmed? Get a goat. Want your siding removed? You need a goat. While goats *can* and *may* consume everything in their path, that doesn't mean they *should*.

Goats are browsers, not grazers. Grazers forage closer to the ground and browsers nibble on stems and branches off the ground. Unlike sheep and cows, goats don't typically graze on grass unless there is nothing else available to eat. Goats prefer brush, limbs, roughage, tree bark, and probably your favorite flower bushes. Free-roaming goats will go from bush to shrub to limb, nibbling a little at each stop before moving on. They browse for tasty treats all day long. In other words, goats need variety for a healthy diet—that's what we'll explore in this chapter.

➥ *Water and Your Goats*

Goats require one to three gallons of clean water per goat per day, and they should always have access to fresh, clean water. They will drink less when there is plenty of fresh green forage and more when they are lactating. It's important to clean out their water buckets and check on them daily. The area around their water should be kept clean and dry, as they stand there often to drink. My goats won't even touch their water if a leaf or a piece of hay falls in. Fickle—and spoiled—goats!

MAIN TYPES OF FEED

Foraging and browsing should always be a goat's main source of food. Hay, grains, and loose minerals make up the remainder of their diet. When fresh forage is unavailable, you need to supplement their diet more with hay and grain.

Forage and hay make up the main portion of a goat's diet, while whole grains make up a small portion. An ideal situation is 75 percent forage, 15 percent hay, and 10 percent grains and minerals.

Always offer your goats free-choice hay—meaning they have access to it 24/7 and are allowed to eat as much as they want. I recommend doing the same with their minerals; place them in a separate feed container available to your goats at all times. Grain should be rationed out and monitored, as goats easily overindulge in grains, which can lead to weight gain and cause health issues.

When storing goat feed, keep it in a dry, covered area with air circulation and free from any contamination, such as mice feces or other animal excrement. Now let's take a look at what's on the menu.

Chaffhaye: This is a fresh, chopped, premium bagged forage. It is harvested at optimal nutritional content, then immediately packaged and sealed green. Chaffhaye is more expensive than the other feed options; however, it has a greater nutritional value. Goats eat less of this than hay, so it will last longer. Some goat owners mix a little Chaffhaye in with their straw to help stretch their supply and save money.

Alfalfa: This is very rich fodder with a higher protein content than hay. This makes it great for goats in milk and young goats. Alfalfa is not suggested for bucks or wethers as they require less protein in their diet.

Straw: This has very little nutritional content for goats and should not be used as feed unless absolutely necessary. It is okay to use for bedding, but a diet of only straw will cause nutritional deficiencies in your goats.

Grain: Only a small portion of a goat's diet should be grain. Many goat owners prefer whole (unprocessed) or rolled grains, such as corn, oats, rye, barley, etc. Processed grains are not recommended for goats.

Sweet grain: This is regular grain with the addition of molasses. Goats can become addicted to sweet feed and will do pretty much anything to get their mouths on it. Sweet grain is an excellent tool for training your goats. All I have to do is put a little sweet grain in a cup and shake it and my goats come running.

Minerals: Think of minerals for your goat like you would think of a multivitamin for yourself. While most of their minerals and nutrition should come from their forage and hay, minerals should be offered to your goats as a free choice. Minerals can be purchased in a loose granular form or in a block, like a salt lick.

Hay: Goats require two to four pounds of hay per day per goat. Hay is sold first, second, and third cutting. First cutting is generally a combination of old growth from winter and new growth. Second cutting is the most desired, as it has fewer weeds and is harvested during prime growing season. Third cutting is in areas with a long growing season and offers a high nutritional value. When buying, ask about the cutting and what types of chemicals or sprays were used on the hay. Next, inspect it. Hay should smell sweet or earthy, never musty or like mold. Moldy hay can cause respiratory and health issues, and old hay lacks nutrition. It should be fresh and pliable; brittle, dry hay is generally too old. Your goats are a good indicator of quality. If they won't eat it, use it for bedding or toss it out.

SETTING A FEEDING SCHEDULE

Like many other animals, goats are creatures of habit. If you start feeding your goats at 8:00 a.m. every day and then decide to sleep in one day, at 8:01 a.m., they are going to let you know you're late. However, it is important to never overfeed your goats. Giving them too much food at once can cause stomach upset and bloat (see page 74).

We feed our goats at 8:00 every morning, giving them half a cup of grain each per goat. They also get fresh hay, free-choice baking soda (to help with pH balance and digestion), free-choice minerals, and several gallons of fresh water. We give them more fresh water in the afternoon or as needed.

Many owners of dairy goats wait until their goats are on the milk stand to feed each one individually. This helps keep the goats occupied during milking and ensures each goat receives the same amount of feed and no single goat is eating more than its fair share. However, you should only feed grain individually; all other feed should be offered free choice.

SPECIAL CONSIDERATIONS

Baby goats, pregnant does, and goats in lactation need more nutrition than other goats. Pregnant goats can eat up to one pound of grain per day, and alfalfa can be added to their hay through their high-lactation days or their first few months of producing milk.

Kids or baby goats won't start eating feed or begin browsing until two to four weeks. They will continue to nurse for their main source of nutrition and food but will begin browsing and tasting the grain. Their increased grain requirements will last until they reach maturity weight.

Remember that goats are what they eat. What your goats eat will affect their milk and meat. If your goats are foraging on wild garlic, wild onions, ivy, mint, or even cabbage, this could alter the taste of their milk. Meat goats that have no access or poor access to quality roughage and forage will tend to have poorer-quality meat.

chapter five

GETTING TO KNOW YOUR KIDS

oats are very lovable, personable, and somewhat unpredictable livestock. They are highly intelligent and very playful. Part of their intelligence is being stubborn. Although they love human interaction, their life goal isn't to please humans. If they don't want to do something, then they are sure to let you know. Ever hear the phrase "stubborn old goat"? Let's review typical goat behavior and identify patterns and signs that reveal underlying things that are affecting your goats.

"NORMAL" GOAT BEHAVIOR

Much like chickens, goats establish a pecking order of sorts. A pecking order is where you have a boss that is at the top, then an assistant manager, the workers, and so on. In some herds, the pecking order remains the same, while in others it may change. This hierarchy helps to establish and maintain balance in the herd.

The goat at the top of the pecking order has a responsibility to its herd members. Each goat has their place in the herd, and every time a new goat is introduced, they too will have to bid for their place in that order.

Some characteristics that determine dominance and hierarchy are age, sex, and horns. For example, does always dominate their female kids no matter how old they get, and goats with horns typically dominate goats without horns.

FINDING A PLACE IN THE HERD

Generally, there are two herd leaders: the top doe and the top buck. The top doe is called the herd queen, and the top buck is called the herd sire. If you are raising your bucks and does separately, the herd queen will be at the top of the pecking order for the female goats. If your bucks and does are together, the herd sire is at the top, then the queen, then the rest of the herd is under them.

In goats' natural habitat, a herd queen leads her herd to find food. She is the one who determines if a plant is poisonous and informs the rest of her herd that they should avoid that plant. The queen is also first in the lunch line and claims the nicest place to sleep. If bucks are not in the herd, the queen is the herd's defender against attacks. The herd sire is the protector of his followers, and he's the first to get in line for the ladies come mating season. He reigns in his position until he is challenged and defeated.

Goats do not see humans as humans. They see humans as funny-looking members of their herd. If you are the one who feeds your goats, they consider you a herd queen. Herd queens and sires will also look at you as a threat to their role and try to dominate you. It is your job and responsibility to show them who's boss. If you push a buck's head or pet his head, he sees that as an invitation to compete for the herd sire position and will try to buck or butt you, so pushing a buck's head is generally not advised.

My husband used to do this until one day when our buck saw him with his back turned. He knocked my husband's feet from under him and he landed flat in the mud! We later corrected this bad behavior (see "Essential Goat Training," page 50), but the look on my muddy husband's face was priceless.

COMMON BEHAVIOR ISSUES

Goats are very playful creatures. They can entertain for hours on end, from jumping and hopping to climbing and playing hide-and-seek. Some of the behavior you may witness as a new goat owner may startle, concern, surprise, or even worry you. Learning what is normal and what isn't will help you to determine if you need to call the vet or just let goats be goats.

Butting heads: Goats butt heads when they are young as a part of playtime. This is practice for when they need to establish dominance as adults. When adults butt heads, they continue until one submits and one wins. If you try to prevent this behavior, they will just pick up where they left off when they get back together. It's just a part of establishing pecking order, so it's best not to interfere.

Climbing and jumping: Goats love to climb, the higher the better. If there is a way to get on top of something—a rooftop, a car top, a tree, a playhouse, an air-conditioner unit, a ladder—your goats will find it. Our small goats can jump amazingly high for their stature. They showed me where our goat-proofing (see page 21) was weak in no time. They also jump incredibly high when spooked. Our spooked Nigerian doe cleared our four-foot fence once with room to spare.

Needing a tissue: Your goat doesn't have a cold, they just like to sneeze. Goats sneeze to alert other goats to danger. Of course, this perceived danger may be a leaf falling, but it's a danger to them nonetheless.

Goats in the moment: A doe in heat or a buck in rut (the male version of heat) will be incorrigible, seriously annoying. A doe in heat without a buck sounds like a young child being tortured in your yard. She will bleat (cry out loud), swish her tail, and want constant attention. The buck in rut will do some crazy things that may seem out of the norm. The only normal thing

about a buck in rut is that he acts abnormally. Bucks in the mood will pee on themselves, creating a sticky residue of urine on their coat. This drives the ladies *crazy*! They lick their urine and make some funny lip-flapping noises. They do not leave the ladies alone and will become more aggressive. You'll learn more about breeding in chapter 7 (see page 65).

ESSENTIAL GOAT TRAINING

Goats are personable and trainable to an extent, much like domestic dogs, but they are still livestock. They are creatures of habit, so when you're training them, repetition and patience are key, as is showing them who's boss. Even if you plan on letting your goats free-roam over hundreds of fenced-in acres, you still want to train them for certain situations, such as moving them, getting them on the milk stand, trimming their hooves (see page 56), and vet checks.

LEASH-TRAINING

Some people use collars on their goats to help with leash-training. Leash-training a goat is done similarly to leash-training dogs. Start working with collars and leashes when they are young so they are used to it as they age. Here's what you can do:

Troubleshooting:
If you plan on using the leash to rotate your herd around your property so the goats have access to different forage, tether the leash to a tree with plenty of slack for twenty minutes at a time. Of course, you will need to stay close by and keep a watchful eye.

Another use for short leashes is to hook goats to a feeding station. This allows the goat to get their fair share of food and/or supplements when you're feeding a lot of goats at one time.

1. Begin with a short leash and practice walking in the yard daily.
2. Hold them close to you and practice leading and stopping.
3. Practice leading young does onto the milk stand, then give them a little treat of food. Once they are comfortable with this, practice making them stay for a short time at first, then increasing the time until they stand still for the length of a full milking session.

chapter six

ROUTINE CARE AND TASKS

In the world of livestock, goats, in my opinion, are some of the easiest to care for. Given the right conditions and the right feed, your healthy goats may seem pretty much self-sufficient. But however easy they may be to raise, they still require care. In this chapter you will learn what care is required for raising healthy and happy goats, from trimming their hooves to castration.

GROOMING

We brush our teeth and hair, take baths, and eat right, so why should our goats be any different? There are two different types of grooming. One is grooming for health and wellness. Routine grooming will help you as a goat owner develop a bond with your goats and allow time for health inspections. The second one is grooming for shows. Obviously, grooming for shows is a lot more involved—regular brushing, bathing, and shaving—since they need to look their absolute best. This section focuses primarily on everyday grooming.

➡ *Petting Your Goat*

Your goats will likely beg you to give them a good scratch just like Fido. Just remember, goats don't like being petted on their head. They get spooked when you try to pet them on the head because they can't see what you're doing. Try petting your goat on the back, chest, or neck. In addition to avoiding the top of their head, try not to push against their forehead. While it's funny and cute when they are little, they are training to knock you off your feet when they get big enough to establish dominance.

BRUSHING AND BATHING

Goats love to get a good back scratch. You will see them brushing up against a tree trunk, a wire fence, or the side of a barn to get that itch scratched. We like to take the heads off of stiff-bristle brooms and screw them to trees so the goats can brush up against them, and they *love* it.

When you spend time brushing your goats, you create a grooming bond, similar to the one a mother makes while grooming her young. You can also take this time to inspect your goat for injuries, abnormalities, and hooves in need of trimming.

During the colder months, goats grow a winter coat. You will see what looks like dryer lint close to their skin. This is their winter fluff. Don't try to brush this out when it's still cold, as they need it to keep them warm. You can, however, help groom them in the spring when they start to shed their winter coat.

To brush your goats, you need a stiff brush and a soft brush. The stiff brush helps get all the old winter fluff off, while the soft brush is for daily or weekly brushing. Brush their fur in the direction it is growing, and try to brush their chest, back, and legs.

If you are raising goats for livestock only, there is little need for bathing unless a goat is sick or gets stuck in the mud, or you just want a better-smelling goat. I recommend reserving the task only for times when it is absolutely necessary. Too much bathing can interfere with the natural oils that keep their skin and coat healthy.

Remember, goats hate water. Bathing your goat is not a fun task for either you or your goat. However, they are creatures of habit, and if you start bathing early, they will get used to it.

If you do want to bathe your goat, I suggest using a collar with a short leash and tethering them to a fence or a milk stand (stanchion) for this task. This holds them still while you remove dirt and shampoo them.

Bucks smell worse than does due to the buck "cologne" they create during mating season. This is a sticky, smelly residue that is quite hard to get off. You actually want the cologne to remain during mating season since the ladies go wild for it. The buck smell does dissipate after mating season has ended.

Use goat milk soap or a livestock shampoo sold at farm supply stores. Wash your goat on a warm day when they will have plenty of time to dry in the sun before nightfall. Avoid cold, wet, or windy days for bathing.

HOOF CARE

As daunting as it may seem, I recommend that all goat owners learn to trim their herd's hooves. Of course, you can hire this job out to trained people, and that may be the best option for you. However, this is a task I am confident you can learn to do on your own with practice.

When a goat lives in its natural environment, rocks, forage, tree bark, and the like keep their hooves trimmed. When you place a goat in a fenced-in pasture, performing routine tasks to care for their hooves is an essential part of protecting their health. Overgrown hooves can lead to leg, joint, and muscle problems as well as foot rot, which happens when bacteria are trapped in the fold on the hoof.

Think of a goat's hooves as you would your fingernails. The growth past the skin is what you need to keep trimmed and clean. Goats have cloven hooves and one dewclaw (on the back of their ankle). How often you will have to trim your goats' hooves depends on the individual goats and their living conditions. It is best to check your goats' hooves weekly to help determine how fast they grow. A general rule of thumb is to trim them every two to four weeks.

When you purchase your goats, ask the breeder to show you how to trim hooves on one of their goats. Take pictures of or notes about the process so you can refer back to it as needed. If your breeder is unable to demonstrate, you can contact your goat vet to help you through the process until you are comfortable in trimming them yourself. Goats are very skittish and jumpy. The last thing you want to do is cut your goat or yourself. Be calm, talk to your goat through the whole process, give it time, and be patient. Do some practice rounds before performing the real deal by walking your goat to the stanchion, securing them, lifting each hoof and trimming it, and then returning them.

HOW TO TRIM YOUR GOAT'S HOOVES

Skill Level: Beginner | **Estimated Material Cost:** $10 or less |
Time: 10 minutes per goat

SUPPLIES AND TOOLS

- Cup of warm water
- Small stiff-bristle brush
- Stanchion or a collar and leash
- Feed or treats, as needed
- Stool or bench
- Clippers or hoof trimmers
- Antiseptic spray
- Styptic spray or powder

Troubleshooting:
Not cutting enough or cutting too much can affect the way the goat walks and can cause issues. It's important to make sure their hooves are trimmed nice and level.

1. If the goat's hooves are muddy, soak them in the warm water and use the brush to remove any mud.
2. Secure the goat to the stanchion. Offer the feed or treats to keep them occupied during trimming. Place the stool beside the goat (not behind), working on just one hoof at a time.
3. Using the warm water and brush, remove any dirt and clean the hooves.
4. Spray the clippers with antiseptic spray. Do this between each trimming to prevent spreading infection or disease from hoof to hoof.
5. Grab the hoof and bend it back toward you; do not raise the leg forward to trim.
6. Using your clippers, trim any excess growth away from the pad of the hoof. The pad is soft and pliable, and the outer edge is the part you will be trimming. Make sure the hoof is level and straight. Any curves should be trimmed.
7. If you accidentally cut the goat, use the antiseptic spray and the styptic spray or powder to stop the bleeding.
8. Dewclaws don't need to be trimmed often, usually only on older goats or when the dewclaws start to curl into the goat's skin.

TO DEHORN OR NOT TO DEHORN, THAT IS THE QUESTION

Dehorning is the act of removing a goat's horns permanently. Dehorning goats is the subject of much debate among goat owners, and both sides have compelling arguments. I have been, and I still am, on both sides. I have goats with horns, goats that have been dehorned, and naturally polled goats, and I've even owned a goat with scurs—we'll get to that in a minute. Let's start with the basic pros and cons of horns.

HORN PROS

➤ Horns help goats defend themselves against predators and protect the herd. This ability is diminished in dehorned goats.

➤ Horns help goats establish a natural pecking order. A goat with horns will dominate a goat without horns.

➤ Horns help regulate body temperature. There are some goats, such as Angora goats, that should never be dehorned because without their horns they could overheat and die.

➤ You avoid the risks associated with dehorning. If not done properly, dehorning can cause permanent brain damage or infection.

HORN CONS

➤ Horns can get stuck in fences and cause injury. This risk is diminished in dehorned goats.

➤ Horns can injure you and others. Dehorned goats are less dangerous to humans.

➤ Goats with horns are often harder to sell than dehorned goats.

➤ Goats with horns are often not allowed in shows. If you plan on showing goats in the future, your goat needs to be dehorned.

Dehorning, or disbudding, is usually done when the goat is just a couple of weeks old. This is a medical procedure in which a hot iron is used to burn the horn buds (horn buds are the beginning of the horns forming) off their head. If you want a goat without horns, let your breeder know before you purchase your goat or contact your vet to perform the procedure.

Polled goats are naturally hornless. A polled goat comes from a parent or parents that are polled. No disbudding or dehorning will ever be needed if you have a polled goat.

Sometimes dehorned goats develop scurs, or partial horns that grow after a goat has been disbudded or dehorned. This can happen months or even years later. Generally, these scurs break off on their own, but sometimes they will continue to grow, which is perfectly fine. I don't recommend trying to remove the scurs at an older age unless it is causing a health issue. If that is the case, it's time to contact a vet.

CASTRATING BUCKS

Many goat owners choose to castrate (remove the testes from) their bucks to control unwanted pregnancies. Other reasons may include genetics or dominance control; herds only need one captain at the helm. Sometimes owners love their bucks but not all the buck traits that go with puberty and rut, so they would rather own a wether.

One buck, or herd sire, can service (perform "the deed" with) anywhere from 10 to 40 does per month, depending on the buck's age. That means you only need one or two intact bucks, depending on your herd size. Any more than that could create a hostile living environment for your girls.

Goats can breed as young as seven weeks old, so it's important to make the decision to castrate or not to castrate before this age.

It is less traumatic and painful for a young buck to be castrated than an older one. Castration should be done at anywhere from three days to three weeks old. Some suggest waiting until the buck is older, following the theory that this helps his bladder system to fully develop, but there is no science to support this idea.

In my opinion, the easiest way to castrate a buck is by banding him using an elastrator, a device that removes the testes by cutting off circulation to the scrotum and testicles. There is no blood, knives, or cutting involved. After 10 to 14 days, the sac shrivels and falls off.

HOW TO CASTRATE

Skill Level: Beginner | **Estimated Material Cost:** $15 to $60 | **Time:** 20 minutes

---

SUPPLIES AND TOOLS

- Elastrator

Troubleshooting:
It is important that both testicles are in the scrotum when positioning the ring close to the body. A scrotum without testicles will feel like an empty skin sack. You should be able to feel two lumps (the testes) inside. If you do not get both testes in the elastrator, the goat will remain fertile.

1. Place the rubber ring onto the prongs of the elastrator.
2. Squeeze the elastrator, which opens the rubber ring. Place the scrotum and testes through the ring, positioning the ring close to the body.
3. Once the placement is secured, slip the rubber ring off of the elastrator prongs.
4. When the testes start falling off, spray antiseptic to avoid infection. If you notice redness or swelling, or the testes have not fallen off after 10 to 14 days, contact your veterinarian.

MICROCHIPPING AND TATTOOING

Microchipping and tattooing your herd help you track your goats in case they get loose or lost. You also need to do this if you want a registered herd (see page 14). Most, if not all, registries require your herd to be tattooed with your own specific numbers and in sequence.

Tattoos are generally placed inside the right or left ear. In some cases, like with LaMancha goats, which don't have extended ears, tattoos are placed under the tail. It is a fairly easy procedure, similar to pressing a seal on paper. You can purchase goat tattoo kits online or at feed supply stores for less than $100.

Microchipping your herd can be quite costly, depending on its size. A local vet can microchip your goats for you. Microchipping usually costs $25 or more per goat.

chapter seven

GOAT HEALTH AND BREEDING

This chapter reviews the fundamentals of goat health, breeding cycles, parasites, and first aid needs. This information is essential. Goats are unlike any other animal. If you are new to raising goats, this information can make the difference between your goats' living and dying. When we were starting out, for example, I was unaware of parasites. Three of my baby goats seemed to be dying, nothing I did helped, and the vet misdiagnosed them because they didn't specialize in goats. Friends eventually helped us figure out the problem and find a good vet, but it just goes to show you that not all animals are the same and it pays to understand a goat's special health needs.

SIGNS THAT YOUR GOAT IS SICK

As a goat owner, you are their first line of defense—the health of your goats largely depends on your ability to diagnose them. Although it's always best to take sick goats to a vet who specializes in working with them, you will be the one to notice the first signs of illness. It's good to have a working knowledge of health issues and treatments in case of emergency, when a vet is not available to you and you need to act fast.

Generally speaking, when goats show signs of sickness, they tend to go downhill fast. Learning to identify the signs of a sick goat will help you to save their life. The first step in identifying a sick goat is to know what a healthy one looks like—refresh your knowledge of a goat's anatomy (see page 10) and how to perform a basic health evaluation (see page 16). You should also spend time with your goat herd, so you have a good sense of their personalities, routine, and normal behavior. That will make changes easier to spot. Last, take some time to familiarize yourself with seven key areas that show signs of illness: eyes, stool, ears, energy, sounds, hair, and temperature.

Eyes: You can tell a lot about a goat's health by the color of the mucous membranes of their eyes. A goat's mucous membranes should be bright pink, almost a hot-pink color. A faded pink or even white color can indicate anemia. The FAMACHA (short for "Faffa Malan Chart"), which you can order online, is used to help identify anemia in goats and sheep. It looks like a set of paint swatches from the hardware store, with shades from pink (healthy) to white (anemia) on it. Anemia in goats is often a sign of parasites (see page 70).

Stool: Goat poop, pellets, or berries—whatever you call it, the stool of a goat is always a good indicator of what is going on inside them. Goat pellets should be round/oval in shape, dark brown, solid (not runny), and firm. They should come out individually, not clumped together, very similar to rabbit manure. Clumpy or runny stool (called scours) can be an indication of stress, a change in diet, parasites, sickness, or infection. If you notice clumpy or runny stool, take the goat's temperature (see page 69) and collect a sample to take to the vet for analysis.

Ears: A goat's ears should be warm to the touch. A hot or cold ear could be a sign that a goat's temperature is off and you should check it (see page 69).

Signs of discharge are another indication of sickness. Droopy ears on a goat whose ears are normally erect are an indication of a vitamin deficiency.

Energy: When not sleeping, healthy goats are active and playful. They forage for much of the day and interact with other members of their herd. They are very social animals. When a goat is isolated or lethargic, it's an indication of sickness, such as parasites, enterotoxemia, bloat (see page 74), or physical injury.

Sounds: A goat makes sounds when they are hungry, in heat, or separated from their mom. They also make sounds of digestion when eating. Other than that, they are relatively quiet animals. A goat that is constantly crying or bleating may be sick, such as with TB (see page 73).

Hair: A goat's hair should be shiny and smooth or wavy. Missing hair could be a sign of parasites, lice, mites, or a mineral deficiency.

Temperature: A goat's temperature should be between 101.5°F and 103.5°F. Use a rectal digital thermometer to take an accurate reading (see page 69). A temperature above 103.5°F or below 101.5°F indicates an illness. Most of the illnesses listed on page 70 will be accompanied by a temperature irregularity.

HOW TO TAKE YOUR GOAT'S TEMPERATURE

Skill Level: Beginner | **Estimated Material Cost:** $10 | **Time:** 5 minutes

SUPPLIES AND TOOLS

- Stanchion or a collar and leash
- Gloves
- Digital rectal thermometer
- Rubbing alcohol
- K-Y Jelly or Vaseline

Troubleshooting:
If the weather is hot outside or your goat has been very active, their temperature can be affected. If you get a reading over 103.5°F during this time, wait a couple of hours and retake their temperature.

1. Secure the goat in the stanchion. Have someone assist you in holding the goat still and keeping it calm.
2. Don the gloves to protect yourself and the goat, then sterilize the thermometer with the rubbing alcohol. Add a small amount of K-Y Jelly or Vaseline to the tip of the thermometer.
3. Lift the goat's tail so you have a clear view of its rectum. Insert the thermometer a couple of inches into the rectum.
4. Wait for the beep and slowly remove the thermometer and record the reading. A normal temperature is 101.5°F to 103.5°F.
5. Sterilize the thermometer and clean the rectum if needed.

COMMON HEALTH PROBLEMS

Let's take a look at common health issues that are specific to goats and similar species. Matching the symptom with the disease or sickness will help you accurately communicate your concern to your veterinarian. This information is not intended for you to self-diagnose and treat your sick goat—I always recommend a vet.

PARASITES

The best way to prevent parasite loads is to have a dry lot, a closed herd, proper rotation, and proper nutrition. However, even in the best of conditions, your livestock can come in contact with parasites, and those parasites can become a problem. A dry lot is land that has never had livestock on it. A closed herd is when you don't introduce livestock from other farms to your animals (so no livestock shows, farm tours, mating visits, or new animals). These measures protect your herd from outside sickness and disease.

Goats can get both external and internal parasites. If you've ever had any kind of domestic animal, you are likely familiar with external parasites, like fleas. They live on the skin and the coats of animals. While not always life-threatening, they can be if left untreated—plus they aren't any fun for your goats. The best prevention is early detection during regular grooming (see page 55). Internal parasites, on the other hand, live inside goats. Goats need a happy balance, and when their parasitic load is too high, they start displaying symptoms of sickness.

Goats get parasites from many sources, such as grass, water, their parents, and soil. Honestly, parasites are unavoidable if you have goats. However, healthy goats can resist them more than those with compromised or weakened immune systems. Let's take a look at the parasites goats often deal with.

Barber pole worms: An internal parasite. Barber pole worms (*Haemonchus contortus*) are stomach worms and destroy the lining of a goat's stomach to reach its bloodstream.

Coccidia: An internal parasite. Coccidia (*Eimeria* or *Isospora*) live in the lining of the intestines and destroy a goat's ability to absorb nutrients.

Fleas: An external parasite. Fleas are not species specific; they can jump from dog to goat to cat, and so on. Cat fleas (*Ctenocephalides felis*) and sticktight fleas (*Echidnophaga gallinacea*) frequently infect goats. Flea infestation can lead to anemia and can cause ulcers on the head and ears.

Lice: An external parasite. Sucking lice (*Linognathus stenopsis*) and biting lice (*Bovicola caprae*) can both be troublesome for your goats, but these species don't infect humans. A lice infestation can cause skin irritation, hair loss, anemia, and even death. Lice are a bigger issue in winter than summer. Effective treatment can be purchased from your veterinarian or local feed store.

Liver flukes: An internal parasite. Once ingested, liver flukes (*Fasciola hepatica*) work their way to the liver, where they destroy organ tissue.

Lungworms: An internal parasite. Lungworms (*Dictyocaulus* spp. or *Muellerius capillaris*) infect the lungs and can cause pneumonia along with secondary infections.

Mites: An external parasite. Many types of mites can plague goats, including follicle mites (*Demodex caprae*), scabies mites (*Sarcoptes scabiei*), psoroptic ear mites (*Psoroptes cuniculi*), and chorioptic scab mites (*Chorioptes bovis*). A mite infestation can cause mange, lesions, hair loss, dermatitis, anemia, and weight loss. Treatment of the entire herd is recommended if mites are found, to prevent further infestation.

Ticks: An external parasite. Ticks are not common on goats but they can get them, including American dog ticks (*Dermacentor variabilis*), Gulf Coast ticks (*Amblyomma maculatum*), and Lone Star ticks (*Amblyomma americanum*). Ticks can spread diseases that affect both goats and humans. They like to hang out in the joints, such as the armpits and legs, and are more prevalent during warm weather. Use extreme caution when removing and disposing of ticks since humans can be infected.

Signs of parasites include:

- Anemia (see page 67)
- Bottle jaw (swelling under the skin right below the jawline; most common with barber pole worms)
- Breathing issues (fast or unusual breathing)
- Coughing (often a sign of lungworms)
- Depression
- Diarrhea/scours
- Fever
- Lethargy (inability to stand or walk around)
- Nasal discharge
- Rough coat (nutrient deficiencies cause a lack of luster)
- Weight loss

VIRUSES AND INFECTIONS

Viruses and infections are a little more serious than parasites, and it's important to detect them early. Some viruses and parasites can infect the soil, where they will remain indefinitely. It's important to screen new goats for diseases before allowing them into your herd via a vet health check (see page 80). Let's take a look at common viruses and infections.

Brucellosis: Brucellosis is a bacterial disease caused by *B. melitensis*. Although it is rare in goats in the United States, it's common in other parts of the world and can be transferred to humans. Goats transfer this disease by licking aborted fetuses, afterbirths, newborns, or vaginal discharges, or by consuming contaminated feed. It can also be transferred by inhaling contaminated dust or by using contaminated milking supplies. Symptoms include late abortions, mastitis, and lameness. Brucellosis can be treated with antibiotics by a veterinarian.

Caprine arthritis and encephalitis (CAE): CAE is a contagious viral disease. It spreads through bodily secretions such as blood, feces, milk, and semen. CAE is not a threat to humans but is life-threatening to goats. Complications from CAE include arthritis, mastitis, encephalitis, pneumonia, and chronic wasting. There is no known cure for CAE and it is usually fatal.

Caseous lymphadenitis (CL): CL is a contagious bacterial infection in the lymph nodes of goats. It is caused by a bacterium called *Corynebacterium pseudotuberculosis*. This bacterium can live in the soil and infect goats for years. It can be transferred through direct contact, soil, bedding, feed containers, grooming equipment, and milking equipment. Flies can also spread this disease when moving from one infected goat to another. There are two forms of this disease, one affecting the lymph nodes right under the skin and the other affecting the lymph nodes inside the body. When the lymph nodes under the skin become infected, they appear swollen under the jaw or the udder. The swollen lymph nodes can rupture, causing mass contamination of your land. Signs of internal CL are weight loss in adult goats or the inability to gain weight in younger goats. CL may be treated with antibiotics by a veterinarian.

Johne's disease: This is a gastrointestinal disease that is fatal in goats. It's caused by the MAP (*Mycobacterium avium* subspecies *paratuberculosis*) bacterium. It's contagious, and the infection generally happens during the first months of a kid's life when they swallow the bacterium via water, milk, or feed that has been contaminated by manure from infected animals. Goats can become infected when young and not show symptoms for years. There is no known cure for Johne's, and it can spread to other species.

Tuberculosis (TB): Tuberculosis (*Mycobacterium bovis*) can be found in goats and transferred to humans. Although goats are quite resistant to it, they can become infected. TB causes lesions in the lungs and other organs. A chronic cough, not to be confused with their predator warning cough, may be a sign of TB and should be tested by a licensed professional. Other signs include loss of appetite, reduced milk production, gargled breathing, and declined health. TB should also be suspected when a goat doesn't respond to antibiotics for respiratory issues.

Our best defense—for both humans and goats—against pathogens is a strong immune system. Things that can weaken your goat's immunity are:

- Change in diet
- Giving birth
- Introducing a new animal into the herd
- Moving to a new location
- Overcrowding
- Poor nutrition
- Sickness
- Stress
- Unclean living conditions
- Weather change

FEED PROBLEMS

Many common health issues in goats are related to feeding: they ate too much, ate too little, or ate the wrong thing. Let's take a close look at these problems and how to identify them.

Bloat: This buildup of gas in your goat's digestive system is caused by a sudden change in diet, moldy or uncured hay, weeds, and even milk replacers. Bloat makes the rumen (part of the goat's stomach) unbalanced and traps gas, causing swelling in one side of the goat. Although common, bloat can cause death if left untreated. Goats have a fairly large abdomen, and it looks like they swallowed a football sideways. When a goat is suffering with bloat, their stomach is uneven. One side will protrude more than the other. We offer free-choice baking soda to help reduce gas and prevent bloat. Other treatments for bloat include giving them a third of a cup of peanut oil.

Enterotoxemia: Also called the overeating disease, enterotoxemia is caused by two different types of bacteria: type C and type D. Generally, goats have these bacteria living in their lower intestines not causing any problems at all. However, when a goat overeats or experiences a change in diet, it can increase the growth of the bacteria, which can lead to death in most cases. Signs include lack of interest in food, inability to get comfortable, restlessness, and diarrhea.

Scours: Scours is basically goat diarrhea. It can be caused by a change in diet, too much food/milk, or a parasite load. Dehydration is your main concern here. Try to identify the cause and keep your goat hydrated with goat electrolytes (which can be found at the feed store), Pedialyte, or a similar product. Remove their grain and make sure they have enough hay. Wash and dry their bum with a clean cloth or paper towel. Keep flies away from their rectal area. Scours that persists for days in spite of your home treatment will need to be treated by a vet. If left untreated, scours can cause death.

BREEDING AND PREGNANCY

Goat babies are hands-down the best part about owning goats! If you're planning to breed your herd, this section will show you how to recognize signs of heat in your does, when to breed, how to breed your does if you don't have a buck, and how to prepare for the kids.

BREEDING SEASON

Some goats are seasonal breeders, meaning they only come into heat during a specific time of year, typically between late summer and late fall like the Alpines, LaManchas, and Nubians. Other breeds, such as Boer, Kiko, and Nigerian, are year-round breeders and can get pregnant anytime during the year. A doe's estrous cycle lasts between 18 and 21 days during the breeding season, and peak fertility varies from a couple of hours to three days.

There are a few signs that your doe is coming into heat. She'll like to bleat, yell, or chat a lot for no obvious reason. She'll swish her tail. You may notice discharge, a wet tail, or swollen vulva. She may act out of character, such as being needy or aggressive—or even mounting other goats.

If you own a buck or herd sire, you will want to give him access to your ladies for two full heat cycles during breeding season to make sure the deed gets done. The average time is 40 to 45 days of exposure in order to ensure pregnancy. However, this is nature, so there are no guarantees.

BUCK SERVICE

If you don't own a buck but want to sell, raise, or milk goats (see chapter 8), you can hire out buck service. It's the same thing as "stud service" in the dog-breeding industry. Many goat association member lists show which breeders offer buck service in your area. Make sure to line up buck service at least a month or more before you need it; these things can't be rushed.

Most breeders who offer buck service will require you to have testing and a wellness check (see page 17) with proof within the last six months, and the breeder should provide you with the same from their buck. You'll also want to make sure the breeder has all the necessary registration paperwork on their buck if you want a registered herd (see page 14). Buck service can work a couple of different ways:

Driveway service: This is where you bring your doe to the buck (or vice versa). Generally, both the doe and the buck are on a leash. A couple of sniffs, then—bing, bam, boom—the job is done and you are on your merry way home. This method isn't always a one-hit wonder, so it sometimes requires repeat visits.

Short visit: A breeder may offer to house your doe at their farm with their buck for two heat cycles to ensure a better chance of pregnancy. They may also offer to have their buck come and stay at your place for a short visit.

Artificial insemination (AI): AI is another viable option. There are many vets that offer this service and will come to your farm to perform insemination. This is more expensive than buck service, and the success rate isn't as high. However, you aren't exposing your does to possible parasites or diseases from a goat outside of your herd.

PREGNANCY AND BIRTH

The gestation period for does is about five months (140 to 155 days). When a doe gives birth and produces milk for the first time, it's called her first "freshening." Subsequent pregnancies are numbers that correspond with the number of births the doe has had: first freshening, second freshening, third freshening, and so on. A doe will behave and look different during her first freshening than during subsequent freshenings.

Most does give birth to two kids per delivery. Although not as common, one and three kids are completely normal as well. Four or more is rare.

You will not necessarily be able to tell your goat is pregnant, other than the lack of signs of heat. Short of getting a pregnancy test from a veterinarian, only time will tell for certain if your doe is pregnant. Closer to her expected delivery, your doe's udder will start to fill out.

Another way to estimate when your doe will deliver is to check her ligaments by her tail. If you run your forefinger and thumb down your goat's spine from the midsection to her tail, right before the beginning of her tail you will feel her ligaments. When a doe is getting ready to kid, those ligaments are gone and you can almost touch your fingers together through her skin at the base of her tail. This is a sure sign she is getting ready to deliver. At this stage, it is a good idea to place Mama in the birthing area (see page 29) and give her plenty of food and water.

ASSEMBLING A BIRTHING KIT

The timing of a goat's birth is not always convenient; it could be in the middle of the night, during a storm—you get the idea. That is why I recommend you prepare for the moment by having a birthing kit ready. Most of the time, your goat will deliver just fine on her own and you won't even be the wiser. You'll just check on your goats one day like always and notice a couple of beautiful babies and the mama, doing what goats do. But there's always that small chance you'll need to step in and help. Here's what you'll need:

○ Baby bottles (in case Mama can't nurse)

○ Colostrum (feed for the newborn kids if Mama can't nurse)

○ Cotton balls

○ Digital scale (to weigh your baby goats)

○ Dog crate or playpen (to help hand-feed the kids)

○ Goat journal (to record details of the delivery including sex, date, weight, color, and other observations)

- ○ Heat lamp (to warm cold babies or Mama)

- ○ Iodine (to use on the umbilical cords)

- ○ K-Y Jelly or antiseptic lubricant (for an assisted delivery)

- ○ Milk replacer (feed for the kids if Mama can't nurse)

- ○ Molasses (to rub on babies' gums to perk them up and feed to Mama after birth for a pick-me-up)

- ○ Nasal syringe (to clean out the kids' airways or nasal passages if clogged)

- ○ Paper towels

- ○ Puppy pads (to place under babies to keep them clean and dry)

- ○ Scissors

- ○ Selenium (if kids are deficient; a floppy ear is a sign of selenium deficiency)

- ○ Surgical gloves (for an assisted delivery)

NEW MAMA

Now that your baby goats have arrived, it is important that you make sure they are nursing and healthy. However, be sure to allow the new mom space and privacy. New moms may act like they don't want to nurse or be with their new kids. You can restrain the doe so the kids have the opportunity to nurse and do this several times a day. If she continues to ignore them or tries to hurt them, you may need to remove the babies and bottle-feed them until they are old enough to join the herd.

If you have to bottle-feed your baby goats, you will need colostrum for their first couple days of feeding and goat milk replacer if you can't get your hands on fresh goat milk. Feed your baby 4 ounces of colostrum 4 times a day for the first 2 days. Increase that amount to 6 ounces by day 3, 9 ounces by day 6, and 12 to 15 ounces per feeding by day 10. Keep up with the 4 feedings a day until the kids are 30 days old, at which time you can decrease the feedings to 3 times per day.

Of course, at this point you may be wondering when you can start milking your goat. For more information, refer to chapter 8 (see page 85).

CALLING THE VET

If you are going to start raising goats, it's important to develop a relationship with your new vet. They will be your lifeline to your goat's health and wellness, starting before you even bring them home (see page 17). Even if your goats are healthy, an annual wellness exam, which usually includes routine bloodwork and a stool analysis, is recommended.

And, of course, you should call the vet whenever you suspect your goat is sick or they display any unusual behavior to schedule an appointment, request a house call, or seek guidance. Although you may feel sad or panicked when you realize your goat might be ill, there are a few steps you should take to minimize the impact to the herd and gather essential information. This will help the vet diagnose the problem as quickly as possible. Here's a checklist you can use:

○ **Take a temperature.** This will be the first thing the veterinarian will ask you for when you call.

○ **Isolate them from the herd.** An area that is dry, warm, and away from other goats is ideal. If your goat is sick, you don't want the rest of your herd getting sick. You also don't want your sick goat to get bullied.

○ **Get a stool sample and place it in a zip-top bag.** The vet will want a fresh sample.

○ **Check their eye membranes for anemia.** Report your findings to the vet.

○ **Make sure they are eating and drinking.** If your goat is not eating or drinking, you may have to hand-feed them by bottle or syringe. Note the last time they ate and drank, as well as any changes to their diet or environment.

○ **Note the age and sex of the goat.** These will be some of the first things the vet asks for.

○ **Note any other additional symptoms.** Explain the symptoms you notice and ask for an appointment or guidance. Some vets will make house calls.

FIRST AID KIT ESSENTIALS

If you have goats, or any livestock, you will need a first aid kit. Below is a list of items I suggest you include in your goat kit.

- Anti-bloat treatment
- Antiseptic spray (for disinfecting wounds or hooves while trimming)
- Apple cider vinegar (for preventive health care, add 1 tablespoon per gallon of water)
- Baby bottle (for when you need to feed your goats or baby goats)
- Bag Balm (udder care for does)
- Baking soda
- Blood Stop (to stop any bleeding from cuts or injuries)
- Bucket
- Collars and a leash
- Electrolyte
- Extra water and feed containers
- Eye wash
- Gauze pads
- Heat lamp (for keeping cold goats warm)
- Hoof trimmers
- Hot-water bottle (for keeping cold goats warm)
- Milk replacer (for baby goats in case the mom can't nurse)
- Molasses (to feed a little bit to lethargic goats or goats that won't eat)
- Nutri-Drench
- Paper towels
- Probiotics
- Rubber gloves
- Rubbing alcohol
- Stanchion/ milk stand
- Thermometer (digital rectal)
- Vaseline or K-Y Jelly
- Vet wrap (the best bandage to use on a wound or injury)
- Zip-top plastic bags (for stool samples)

VACCINATIONS

As the saying goes, an ounce of prevention is worth a pound of cure. The main vaccination recommended for all goats is CD & T. CD & T is a vaccination against *Clostridium perfringens* type C + D (enterotoxemia) and *Clostridium tetani* (tetanus).

It is recommended that the initial vaccination be followed by a booster three to four weeks later. Kids should be vaccinated at six to eight weeks of age, with vaccination occurring before weaning. Booster vaccines should be administered to kids three to six weeks later, depending on vet recommendations. Follow-up CD & T vaccines are recommended at least annually or semiannually. The duration of immunity is shorter in goats versus other animals, so immunization is recommended more frequently than with other animals.

CD & T is the only vaccination with a blanket recommendation for goats. All other immunizations should be administered only if there is a local or regional reason for it. This is why it is important to have a good relationship with your veterinarian, so you can follow their recommendations.

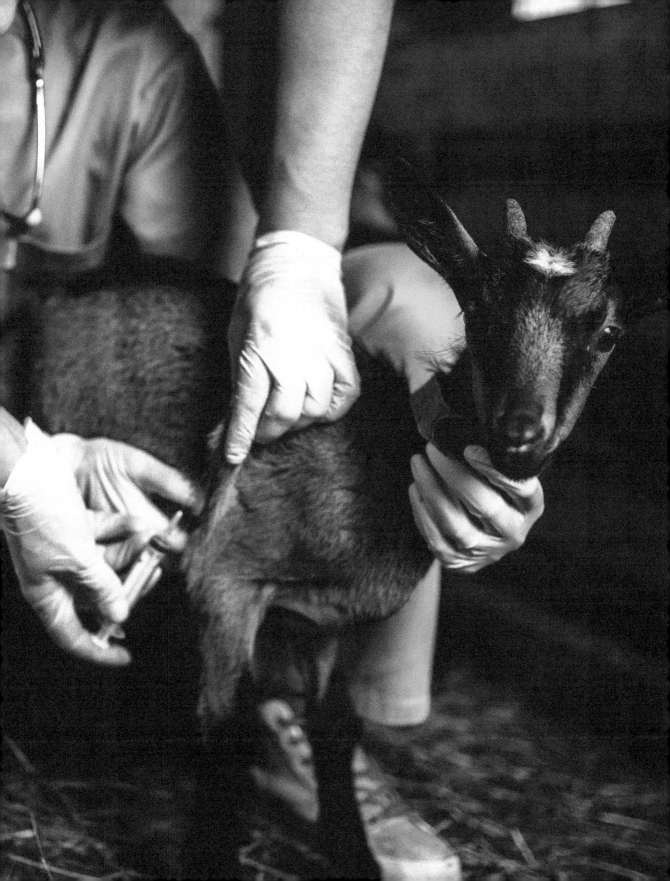

chapter eight

PRODUCING GOAT MILK

oat milk is closer in composition to human milk than cow milk. Compared with cow milk, goat milk is higher in butterfat, lower in cholesterol, and easier to digest, making goat milk a superior choice, in my opinion (of course I may be a bit biased!). In fact, it is the most preferred source of milk in many countries aside from the United States.

Owning your own dairy production brings you one step closer to true self-sufficiency. The more you are able to produce on your homestead, the less you rely on the grocery store. Not only that, but you are now connected with the food source. There is no middleman between you and where your food comes from. You have total control over the production and quality—how empowering is that?

If you've never milked a goat before, or any dairy animal for that matter, the task may seem a bit daunting. This chapter covers goat milk basics and walks you through the production process step by step.

BUDGET BREAKDOWN, EXPECTED REVENUE, AND TIME FRAME

Whether you're running a farm to be self-sustaining or you want to set up a side business, your budget will make or break your homestead. I was not raised on a farm, nor did I have much livestock experience, but as I worked toward my goal of self-sufficiency, I realized goats would be a part of the equation. They could supply milk, and I wouldn't need the acreage (or the pounds of food) required for cattle. It seemed like a quick solution. What I didn't realize is that you need a mama in milk—and that means you need a buck to do the deed, a baby, and the time necessary for Mother Nature to do her magic. We ended up adopting a doe that was in milk and her baby, so we got up and running right away.

Here's a rough budget of the investment you may need to make up front, assuming you are planning to use buck service and have already made the improvements (shelter, fencing, etc.) to your land necessary to raise milk goats and their babies.

➨ Livestock (per doe): $50 to $500 (depending on breed)

➨ Buck service (per doe): $25 to $50

➨ Milking equipment: $150 to $500

If you only plan to keep a doe or two and produce for your family, you may not require a milking machine. But if you plan to scale your operation, a milking machine likely makes more sense.

Depending on the dairy breed you choose (see page 4), one doe can produce 1,200 to 2,600 pounds of milk per year. The average milk price fluctuates, but recently it's been about $5 to $10 per gallon. That means revenues could be potentially $750 to $3,250 per doe per year. However, that does not account for any expenses or labor. I recommend seeking out resources from your local farm extension or university agricultural department for more detailed information about dairy goat operation costs in your area. Both Iowa State University and the University of Wisconsin, for example, have resources and example budgets available online.

In terms of a time frame, expect anywhere from 10 to 15 months for your doe to get pregnant, and then it's about 5 months until the babies are born. However, you can't start milking right away—babies need nutrients from their mama before she can share her milk with you.

Once the kids are two to three weeks and they begin to forage a little on their own, you can start separating the mama from the kids at night and milking your doe in the morning (or vice versa). The kids can be removed from the mom and weaned at eight to ten weeks of age.

This means, if all the stars align, with an adult doe who is pregnant from beginning to end, you could have fresh milk in your cup in six months. You can milk a goat for ten months, then let her dry up for two months before breeding her again.

GOAT MILK BASICS

Dairy goats are productive compared to the amount of food they eat—on average a doe should be given one pound of feed per day for every two quarts (or four pounds) of milk she produces. The rest of a doe's food intake should be from both hay and food found when browsing. If raising goats is part of your homestead, four to seven pounds of food a day per goat may be more than enough for a family's needs, and you may be able to sell the surplus (or products made from the surplus) for a profit.

One thing to keep in mind is that the supply of goat milk is relatively low in the United States because demand is lower. This often means buyers are accustomed to paying higher prices per gallon than they would for cow milk. Aside from goat milk for consumption, you can choose to produce goat cheese, goat yogurt, goat butter, goat ice cream, goat milk fudge, or goat milk soap—or sell to someone who does. There is so much you can do with goat milk to monetize your herd and save money by producing your own dairy needs instead of buying them from a store.

KNOWING WHEN TO MILK

Time is the biggest consideration when deciding to own dairy goats. This isn't something you can decide to do and then change your mind when you have a doe in milk. It's important that you understand the commitment that you will need to make to your girls during milking. In my opinion, the rewards are worth it, but it is a daily commitment and it can't be missed—not for rain, illness, or vacation. I recommend finding a milk buddy who knows how to milk goats—a plan B in the event you can't be home in time for milking.

When it comes to a milking schedule, you have a couple options:

1. You can remove the kids when they get old enough and milk twice a day, resulting in more milk for you.
2. You can raise the kids with the doe so you only have to milk once a day, resulting in less milk for you.

Either way, your doe will have to be milked at least twice a day, by you and/or by her kids. Failure to milk or missing a milking can result in engorged teats and mastitis, which are dangerous and painful for your doe.

Do what works best for you. If you are a morning person, separate the kids and mom at night and milk at sunrise. If you like to do your farm chores at night, separate the kids during the day, milk at night, then place the kids with the mama. Once you get a routine down, it will become second nature to both you and your goats.

KEEPING MILK FRESH

Here is where goat milk gets a bad rap: goat milk that is not processed properly will have that funny taste that people accuse goat milk of having. How you process the milk is the number one factor affecting its taste. For years people have been blaming the goats when in reality it could be the fault of the person who is processing it. Here are some reasons why goat milk might taste off:

➡ The teats weren't cleaned

➡ The milking supplies were dirty

➡ The milk wasn't cooled quickly enough or to the proper temperature

➡ The goat has an infection

➺ The bucks were kept too close to the does—bucks cause the ladies to produce pheromones, and these pheromones can alter the taste of the milk

➺ The milk wasn't filtered

➺ The goat's diet

For more information on best practices for processing and storage, see page 96.

BASIC SUPPLIES

It's best to purchase all your goat-milking supplies before your goats kid. You can get very basic or pretty fancy when it comes to milking supplies. I personally like to keep things simple. Here's my checklist:

○ **Stanchion:** A stanchion (milk stand) is a platform that your goats stand on and are secured to durin milking, grooming, feeding, or health inspections. It secures their head so that they can't injure you or themselves, and it holds them in place. This is your lifesaver when it comes to milking.

○ **Leash and collar:** Use these to walk your goats to the stanchion.

○ **Feed bucket:** Train your goats to feed on the stanchion. This will make getting them on the milk stand a lot easier.

○ **Two stainless steel buckets:** This is what you will milk in. I like to keep mine in the fridge so they are cold for milking.

○ **Fine mesh strainer:** You will need to strain your milk before refrigerating.

○ **Glass quart canning jars with plastic lids:** This is the easiest thing to store your fresh milk in.

○ **Teat spray or wipes:** This cleans the teats before milking.

○ **Strip cup:** Something as simple as a paper cup will work.

○ **Teat dip:** Used to protect the teats between milking.

○ **Milk stool:** This stool is what you will use to sit on during milking. I like having one that is adjustable and comfortable.

○ **Paper towels:** To wipe teats clean after the teat spray.

MILKING YOUR GOATS AND PROCESSING MILK

Goats have two teats and both will need to be milked at each milking. The best advice I have for learning how to milk a goat for the first time is to relax. If you are stressed, upset, and impatient, milking will be a bad experience for both you and your goats.

Owning a milking machine is wonderful if you have multiple does to milk or if you have limited mobility in your hands. It is an amazing invention and saves you a ton of time. Even if you plan to use an electric milker, I strongly suggest you milk by hand first. Machines break, power goes out, things happen, but your goats still need to be milked. If you start out hand-milking, you will always have the knowledge and experience should you ever need to fall back on it.

Whether you use your hands or a machine, you need to strip the teats by hand first. *Stripping the teat* means disposing of the first stream of milk into a "strip cup," which is separate from your milk pail. The first bit of milk is usually not fit for consumption. You can also examine this first bit of milk to check for signs that the milk is off or that your doe is developing a problem (see page 92).

HOW TO HAND-MILK YOUR GOATS

Skill Level: Beginner | **Estimated Material Cost:** $0 to $200 |
Time: 10 to 20 minutes per goat

SUPPLIES AND TOOLS

- **Feed bucket with grain**
- **Leash and collar**
- **Stanchion or milking stand**
- **Teat spray or wipes**
- **Paper towels**
- **Strip cup**
- **Stainless steel milk pail (chilled)**

Troubleshooting:
Make sure you talk to your goat throughout the process to keep her calm. If I don't detect an off color or smell, I like to taste a tiny bit of the milk in step 9, just to be sure.

1. Wash and dry your hands. Fill the feed bucket with the appropriate amount of grain (up to 1 pound per day for goats in milk) and attach to the front of the stanchion.

2. Clip the leash to the goat's collar and lead her to the stanchion. Once the goat is on the milk stand, clip the leash to the stanchion and close the head gate (this holds the goat in place while you're milking).

3. Spray the udder and teats with teat spray or wipes. Dry teats and udder with paper towels.

4. Rub your hands together to warm them up. Sit level with the goat's hooves so you have a clear view of the teats.

5. Grab the teat closest to you. Place your thumb and forefinger close to the udder, then bump the udder up with your hand. The bump mimics what the goat babies do to cause the udder to release milk.

6. Express into the strip cup first so you can test the milk. To do this, put the strip cup into position under the teat.

7. Pinch your thumb and forefinger together and compress the rest of your fingers toward your palm. Try not to lose pressure between your forefinger and thumb. You are trapping the milk in the teat with these two fingers and using the rest of your fingers to squeeze the milk out the end of the teat. You are expressing the milk out of the teats by squeezing them, not pulling them down from the udder, like you would milk a cow.

8. After the milk is expressed, open your hand completely and repeat step 7. You will do this 2 or 3 times per teat to strip it.

9. Examine the color and smell of the milk to detect anything off. Perform this simple test every time you milk your goat to help eliminate bacteria in the teat and to identify issues with the milk.

10. Repeat steps 5 through 9 with the other teat, then discard the strip milk.

11. Place the milk pail under the udder and repeat steps 7 and 8 until you can no longer express milk and the udder feels empty.

12. Bump the udder one more time and try to express milk again. Do this to both teats.

13. Release your goat from the stanchion and bring her back to the yard.

HOW TO MACHINE-MILK YOUR GOATS

Skill Level: Beginner | **Estimated Material Cost:** $150 to $400 |
Time: 10 minutes

SUPPLIES AND TOOLS

- **Feed bucket with grain**
- **Leash and collar**
- **Stanchion or milking stand**
- **Teat spray or wipes**
- **Paper towels**
- **Strip cup**
- **Stainless steel milk pail (chilled)**
- **Milk machine**

1. Wash and dry your hands. Fill the feed bucket with the appropriate amount of grain (up to 1 pound per day for goats in milk) and attach to the front of the stanchion.

2. Clip the leash to the goat's collar and lead her to the stanchion. Once the goat is on the milk stand, clip the leash to the stanchion and close the head gate (this holds the goat in place while you're milking).

3. Spray the udder and teats with teat spray or wipes. Dry teats and udder with paper towels.

4. Rub your hands together to warm them up. Sit level with the goat's hooves so you have a clear view of the teats.

5. Grab the teat closest to you. Place your thumb and forefinger close to the udder, then bump the udder up with your hand. The bump mimics what the goat babies do to cause the udder to release milk.

6. Express into the strip cup first so you can test the milk. To do this, put the strip cup into position under the teat.

7. Pinch your thumb and forefinger together and compress the rest of your fingers toward your palm. Try not to lose pressure between your forefinger and thumb. You are trapping the milk in the teat with these two fingers and using the rest of your fingers to squeeze the milk out the end of the teat. You are expressing the milk out of the teats by squeezing them, not pulling them down from the udder, like you would milk a cow.

8. After the milk is expressed, open your hand completely and repeat step 7. You will do this 2 or 3 times per teat to strip it.

9. Examine the color and smell of the milk to detect anything off. Perform this simple test every time you milk your goat to help eliminate bacteria in the teat and to identify issues with the milk.

10. Repeat steps 5 through 9 with the other teat, then discard the strip milk.

11. Turn on the milk machine and attach the claws (the part that goes on the udder) to each teat and make sure that it has made a good seal. You shouldn't be able to pull it off easily if there is a good connection between claw and teat.

12. When the udder is emptied, use your finger to release the seal of the claw.

13. Release your goat from the stanchion and bring her back to the yard.

14. Clean the milk machine thoroughly per the manufacturer's instructions.

PROCESSING THE MILK

Once you have placed your doe back with the herd, you want to get your milk processed and chilled as quickly as possible. Here are some key points:

➡ Milk should be stored between 35°F and 38°F and maintained at that temperature. Temperatures higher than 38°F can cause goat milk to taste musty or sour. I chill both of my stainless steel buckets in the fridge before milking in order to chill the milk as quickly as possible.

➡ Once you have milked your goat into your first chilled bucket, strain the milk as soon as possible with a milk filter or fine stainless steel filter into your second chilled stainless steel bucket.

➡ Get a chilled glass quart jar and strain your milk a second time from the steel bucket into the glass jar.

➡ Screw the plastic jar lid on top and use an erasable marker to date the milk and include the name of the goat. I do this so if something tastes wrong with the milk, I know which goat it came from.

➡ Immediately store in a fridge or cooler.

There is a lot of debate about the benefits of raw milk and the process of pasteurizing it. Personally, I only drink raw milk as my body does not respond well to pasteurized milk. However, raw milk sales are illegal in many states, so pasteurization in those states is mandatory. Pasteurization can be accomplished by heating the raw milk in a stainless steel pot over medium to high heat until it reaches a temperature of 165°F or higher for fifteen seconds.

TROUBLESHOOTING TIPS

Goats face many of the same complications as humans do with milk production: painful infections, swollen and hard teats, and engorgement from lack of milking. It's important to recognize issues when they start and treat them immediately. If left untreated, udder and teat complications can result in death.

Mastitis: This causes hot, red, and swollen teats. This is an infection and needs to be treated by a veterinarian so that antibiotics can be administered.

Blood in milk: This can be caused by mastitis or an injury to the udder or teats. A slight crack in the skin can cause blood to appear in the milk.

Hot, swollen udder: Sometimes massaging and frequent milking can solve this. It could also mean an infection is present. Start by taking a temperature and call your vet.

RUNNING YOUR SIDE BUSINESS

A dairy goat can be a lucrative way to monetize your homestead—from goat milk production to goat milk products like soap and lotion. Once you are producing enough milk for your own consumption, you may want to do something with the surplus.

Before you decide to invest in dairy goats as a business, investigate your local laws about the legality of selling milk, the requirements, and the licensing that may be needed. Selling milk and milk products is regulated by state and federal governments. In order to sell goat milk or goat milk products, contact your local agriculture department or extension office to find out what you need to do to set up shop. In some states, for example, selling raw goat milk is illegal for human consumption but legal for animal consumption. Other states offer herd-shares, meaning the customer buys a share of your goat, which makes them part owner and legally able to use the milk. Selling goat milk products, such as soap or lotion, may be regulated by your state's cottage laws, laws that allow small producers to create goods at home.

When I lived in South Carolina, selling goat milk for human consumption was legal, but in my new home state of Tennessee, it's only legal to sell it for animal consumption. Always check the laws in your state to protect yourself against any fines or prosecution.

How much money you can make from your milk surplus depends on demand in your local area. I've sold milk for animal consumption for $9 per gallon and goat soap for $5 per single bar. Our small operation sold products at farm sales, food co-ops, mom-and-pop shops, farmers' markets, garden clubs, local artisan stores, and farm centers. If you have two goats in milk that produce a surplus of two gallons of milk per day, that math can quickly add up and turn into a profitable business.

chapter nine

HARVESTING GOAT MEAT

I f you are looking for livestock to raise for meat, raising goats is an excellent option for those with limited land or who want smaller livestock than cows. Goats reach full maturity weight in less time than cows: less than a year for goats, compared with 18 months for cows. This chapter will explore how you can ethically raise and harvest goats for sustainability or business purposes, including ethical slaughtering practices and how to find buyers.

BUDGET BREAKDOWN, EXPECTED REVENUE, AND TIME FRAME

Meat goats reach butcher age between 6 and 9 months of age. Some ethnic groups prefer younger goats for their religious celebrations, so depending on your market, you may be able to harvest sooner. Goats between 40 and 60 pounds sell the best for the meat market, so the growth time is reduced and you have the opportunity to turn a quicker profit.

The USDA provides a wonderful resource called the National Monthly Grass Fed Lamb and Goat Report. It lists what the goat prices are per pound per cut and the average feed cost. This is a good base for you to price your meat and estimate costs and revenue. For example, at the time of this writing, a goat shank was selling for $7.50 per pound and a sirloin was selling for $24.99 per pound. These prices fluctuate monthly but they give you an idea.

A lot will also depend on the supply and demand in your area. If you're the only show in town and everyone wants goat meat, you can charge a higher price than if you were in an area where goat meat is abundant.

Now let's talk about additional investment. You can purchase meat goats for $25 to $100 per goat, depending on the supply and demand. Since one buck can service up to 40 does, you will want to invest in a good-quality herd sire, one with good bone structure and health.

Each doe will produce one to three kids per year, and each kid will grow to butcher weight in under a year. The father's genes determine how fast the kids grow as well as their meat development.

The better the pasture you have to offer your goats, the less you have to spend on feed for them. If you have great pasture, you can raise six to eight meat goats per acre.

I have never met a small-scale or even a medium-scale meat-goat farmer who thought it would be worth the investment to build their own certified meat-processing plant. It costs thousands and thousands of dollars in equipment and licensing. Not to mention all the inspections and insurance.

Since your customer will be paying for the processing, your costs involved in raising meat goats will be your initial investment in your breeding pair, vet bills, feed bills, shelter, fencing, and any advertising. Of course, I've always believed the best advertising is word of mouth. You take good care of your customers, they will take care of you.

GOAT MEAT BASICS

Goat meat is the fourth-most-popular meat consumed in the world, and the market is vastly expanding here in the United States. There are not enough farms producing goat meat to meet the demand, so we import around 50 percent of the goat meat consumed here. The goat meat that is imported is frozen, but many consumers are searching for a fresh supply. Here is where you come in. According to consumer reports, the cost for goat meat has tripled in just the past few years. Cha-ching!

Ethnic groups as well as specialty shops and those seeking a low-fat meat option are consuming more goat meat. Goat meat is lean, with over 50 percent less fat than beef. In addition to being lean, goats are typically raised on pasture or free-range, which dovetails nicely with the movement toward ethically raised meat.

Goat meat is considered a red meat. Goat kids are harvested when they reach 25 to 45 pounds. They are more tender but not as flavorful and juicy as older goats. Ideally you will want to harvest all meat goats before they are a year old, as older goat meat gets tough. The males have less fat than females, but meat from the female goats is more tender, perfect for steaks and chops.

BEST CROSSBREED OPTIONS

Many dairy goat breeds, such as Nubian and Alpine, are crossed with Boers to produce a breed with more desirable meat. However, there are also new crossbreeds available that are produced by crossing different species of meat goats. Some of the new meat-goat crosses include:

➡ **Savanna:** Cross between South African Landraces and Boers

➡ **Moneymaker:** Saanen + Nubian, which is then bred with a Boer

➡ **TexMaster:** Cross between Boers and Tennessee fainting goats

While you can still harvest a dairy goat for meat, they won't have as much meat per pound as the meat breeds. Some of the crossbreeds are sought after by those who want a little more dairy from their meat goats. Many say that the crossbreeds are heartier and more resistant to parasites.

SLAUGHTERING

When you decide to raise goats for meat, you need to decide how you plan on harvesting the animal. When it comes to selling your goat meat, you have to use a processor. However, if you are harvesting the meat for your own consumption, you can do it yourself or hire it out. This decision may change in time. Some goat owners want nothing to do with the killing or processing portion of raising meat goats, then they change their minds after seeing the money they can save.

State and federal governments have laws to help ensure livestock is processed in a humane way. For example, nearly every state in the United States requires that the animal be rendered insensible to pain before processing begins. In other words, it is not humane (or legal) to slit the throat of livestock and let it bleed out. You need to make a kill shot first so the animal is unable to feel any pain.

There are three different ways you can go about harvesting your livestock: you can do it yourself (for personal consumption only), take it to a licensed processor, or hire a mobile butcher.

DOING IT YOURSELF

A true path toward self-sufficiency is doing as much as you can yourself, and this includes harvesting your livestock when it's time. Realistically, this isn't something we all can do. I have a rancher friend who can process her goats with no problem but can't take the kill shot. I know others who can take the kill shot but can't handle the processing. Then there are some who don't have a problem with either.

We are all different, even when it comes to harvesting our livestock. Fortunately, there are options. That said, even if you don't plan on dispatching your goats yourself, I recommend participating in the process at least once. It's good to really understand the full process, but my advice is practical, too. There may be times when a sick or injured goat may need to be quickly and humanely put down and there won't be a vet or butcher available. Knowledge is always good to have, even if you never have to apply it.

Processing meat yourself can save money. The more you can save, the more you can make. Another benefit is time. Butchers generally have a waiting list,

SEVEN POPULAR GOAT CUTS

Goat meat is easy to cook, tender and juicy, and can be prepared in many delicious ways. From braising and stewing, to roasting, grilling, barbecuing and pan-frying, goat meat can taste like lamb or beef. Below you'll find seven popular cuts that are good for different types of chops, tenderloins, racks, cutlets, steaks, ribs, legs, and more.

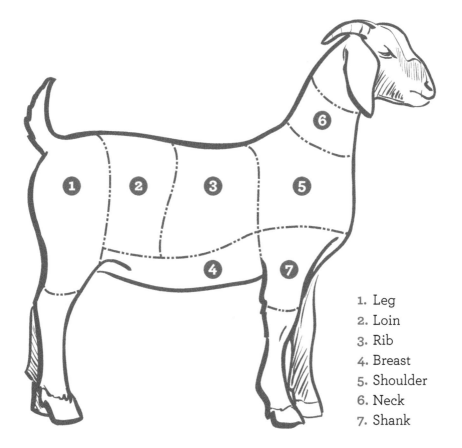

1. Leg
2. Loin
3. Rib
4. Breast
5. Shoulder
6. Neck
7. Shank

and once they do get to you, it takes time to process your meat. If you self-butcher, you get the meat right then and there, no waiting. Another benefit of self-processing is knowing you're getting your own meat. Unfortunately, there are some butchers and processors who don't ensure you are getting your own meat back, so in fact, you may be getting someone else's.

USING A LICENSED SLAUGHTER PLANT

You will need to use a licensed slaughterhouse if you want to sell your meat. A licensed processor can be inspected by the state, the USDA, or both. A good processor is worth their weight in gold. A bad processor is like throwing money in the trash.

Outside of giving you the ability to resell your meat, a good processor can cool your meat properly, store it, and make all the professional cuts and blends you want. Their facilities are clean (or at least they should be) and inspected, and the meat is handled properly in order to reduce contamination and bacteria growth.

Depending on the size of your goat, you can plan to spend $40 or more per goat, and then more for specific cuts and packaging.

HIRING A MOBILE BUTCHER

A mobile butcher comes to your home and processes your animal. It is easier on the animal and causes less stress (stress can negatively affect the meat quality). If you plan on selling your meat, your mobile butcher must be properly licensed and inspected in accordance with federal and local laws. If not, you will not be able to sell it.

A mobile butcher disposes of the carcass and entrails for you. They usually charge a kill fee and a processing fee. Most offer group discounts, so it would save you money if you harvest more than one goat at a time.

RUNNING YOUR SIDE BUSINESS

If you plan to sell goat meat, it's important to market your meat to reach your buyers. You have likely seen products with labels like "organic," "all-natural," and "grass-fed." As the popularity of some of these labels grows, so do the licensing requirements and regulations to use them. You can no longer say something is organic without being a licensed organic farm, which requires fees. If your farm is organic but you don't pay the fee, you can't call your products organic.

However, there are still honest marketing approaches that you can use without having to pay that will help you reach your target customer. For instance, my farm friend who lives out West, where there are a lot of devastating fires, advertises, "These goats were raised to maintain a fire-safe environment," which appeals to her local market.

There are a lot of ways you can market your product to find your own niche and draw in the crowds. Here are some selling points you can think about for marketing your goats:

- Family-farm raised
- Pasture raised
- Free-range
- Humanely processed
- All natural
- Locally produced
- Farm to table

Although goat meat is widely consumed in other parts of the world over beef or poultry, it is fourth on the list in the United States, with the exception of ethnic and specialty markets. Certain ethnic groups—such as those with heritage from Asia, the Middle East, Latin America, Africa, and the Caribbean—will likely be your largest consumers.

Religious celebrations where goat is on the menu for celebratory or ceremonial meals often encourage a spike in demand. Thankfully, there are many celebrations throughout the year, giving your business a year-round market. A few of these holidays include:

- Chinese New Year
- Greek Orthodox Easter
- Rosh Hashanah
- Islamic New Year
- Start of Ramadan
- Passover

To find your customer base, you need to go where they are. Start by advertising at specialty stores, farmers' markets, local restaurants, religious centers, and even livestock auctions. Create a desire in people you already know by inviting them over for dinner and letting them see how good the goat tastes. Once people see and experience the difference in farm-raised meat that is treated ethically and humanely, you can create your own market. Help educate those in your community about the benefits of goat meat, how lean it is, how sustainable the livestock is, and the importance of cutting out the middleman and getting their meat from the source.

Keep in mind that on-the-farm slaughtering for sale is illegal and you can only sell the live animal—what we call "on the hoof." The purchaser can then transport the animal to the slaughterhouse or have a mobile butcher process it.

LEGAL REQUIREMENTS

Every state has specific laws about selling meat, and then there are also federal laws. A good place to start in learning these laws is to contact your local state extension agency. You can also contact a local USDA inspector and ask for information. In order to be in compliance, you need to follow both state and federal regulations. I would strongly suggest you find out about all your local laws pertaining to selling meat before you buy your goats.

It is illegal to process meat on your land and sell it. All meat sales have to be processed at a licensed federal or state inspected processing facility. In order to find a licensed facility, you can contact your local extension agency or goat registry, or ask your local butcher.

chapter ten

ADDITIONAL GOAT BENEFITS

If you are thinking about sustainable living or looking for a way to make money off of your homestead's livestock, you can't do better than goats. Simply put, they are amazingly versatile. No other animal on our homestead is as multifaceted.

In order to become truly self-sufficient, you need to be able to produce what your family needs and make an income. Let's face it: we will always have taxes and we will always have things we need cash for. And goats can help you achieve that—in time, of course.

There are many ways you can monetize your goat herd. Let's take a closer look at additional benefits like selling fiber, compost, and weed control.

SELLING FIBER

When we think of fiber animals, the first thing that comes to mind is sheep. Although sheep's wool may be more popular, fiber from goats is considered more luxurious and extravagant. Their fiber is soft, durable, long lasting, and used in the finest garments.

Selling fiber from your goats is just one of the many ways you can use goats to help sustain your self-sufficient life and make money from home. In addition to monetizing your herd by selling fibers, you can also save money by making your own clothes from your goat fiber. Try your hand at using a spindle and loom, or even knitting. You can make some beautiful clothes from your goat fibers. Handmade items are also in high demand at artisan stores and markets.

The average cost for wool in the United States is $1.75 per pound, whereas fiber from goats can range from $40 to $190 per pound. Some sheep farmers have turned to raising sheep for meat and raising goats for fiber to make more money.

FIBER TYPES

There are three different types of main fibers you get from goats: mohair, cashmere, and cashgora. Let's take a look:

Mohair is the fiber most used in the goat textile industry and comes from Angora goats: It is a strong, beautiful fiber. Turkey was the original producer of mohair, but the demand exceeded their supply and now the United States is the biggest producer of mohair.

Cashmere is the most luxurious textile and highly sought after. Cashmere also commands higher prices due to its qualities. Cashmere is produced by many goats, including Pygora and Nigora goats.

Cashgora is a happy result of cross-breeding mohair and cashmere goats. Because of this cross-breeding, not every cashgora goat is able to produce high-quality fiber. Sometimes it's the luck of the draw. Because of this lack of consistency in breed quality, cashgora is considered rare and in demand.

For more information on fiber breeds, see page 8.

SHEARING

When raising fiber goats, you have one of two choices when it comes to shearing time: do it yourself or hire out the work. Hiring out the task, of course, costs more money. Depending on the type of goat and where you live, you might expect to pay an average of $16 to $25 per goat.

But learning how to shear well takes practice. It's not something to be done in a hurry and you have to have extreme patience. A well-trained shearer can accomplish this task in minutes, while a novice shearer can take one to three hours, which can be stressful on the goat.

Mohair goats need to be shorn twice a year, like sheep, to get their fiber. However, cashmere goats need to be brushed to get their fiber, making the process more time-consuming.

If you are planning on shearing your own herd, I suggest you hire a professional the first time and take really good notes. This way you can observe the process and the techniques they use. In the meantime, here are some general tips:

➻ Expect to harvest five to ten pounds of fiber per adult goat per shearing.

➻ Angora goats should be shorn in the spring right before kidding season and in the fall before mating season.

➻ Goats can be sheared when their hair reaches four to six inches in length.

➻ Have good lighting and your first aid kit handy, including the Blood Stop.

➻ Goats need to be cleaned and completely dry before shearing.

➻ You can use shearing scissors or an electric livestock shear (one designed for sheep).

➻ Place a drop cloth or tarp down, then your stanchion on top of it to keep your goat in place while shearing.

➻ Work on one side of your goat at a time.

➻ Shave as close to the skin as possible without cutting the skin.

➺ Have two catchment containers available, one for the discard hair, and one for the fiber you want to process. The discard hair is any coarse or undesirable hair. Use a breathable bag for the fiber you want to process.

➺ Label and tag your bag with the date and the age, weight, and name of the goat.

PROCESSING

Once you have fiber from your goats, it has to be made into something for resale. You can sell raw fiber, as many new homesteaders want to do the processing themselves, or you can sell the ready-to-use product. The first step in processing is washing. You need to wash the hair to remove any dirt, grease, or impurities. You can wash your goat fiber in the washing machine, but I would recommend doing it by hand.

HOW TO WASH GOAT FIBER BY HAND

Skill Level: Beginner | **Estimated Material Cost:** $20 |
Time: At least 30 minutes

SUPPLIES AND TOOLS

- Cloth mesh wash bags
- Washtub
- Hot water
- Detergent or Dawn dish soap
- Drying rack

Troubleshooting:
Be careful not to wring the fiber or use the agitator in the washing machine. Doing so will cause the fibers to felt, which isn't good.

1. Place the fibers loosely in the wash bag. Do not pack it tight.
2. Fill the tub with water between 143°F and 160°F.
3. Add laundry detergent (a quarter cup per pound of fiber) to the water and mix.
4. Gently add the bag of fibers to the water and soak for a couple of minutes.
5. Squeeze out the water.
6. Repeat steps 2 through 5.
7. Once the fiber is clean (dirt-free and bright), you will need to rinse the soap out following the same process in steps 2 through 5 (without detergent).
8. Lay the mesh bag on a drying rack to continue draining the water.
9. Once the water is drained, you can remove the fiber and place it on the drying rack until completely dry.

SPINNING

After you wash the fiber, you card or comb it. Carding and combing combine the fibers together, straighten them out, and remove any debris that the washing missed, so that the fiber is ready to be spun into yarn.

To spin fiber into yarn you can use a drop spindle or a spinning wheel. A drop spindle looks like a stick the size of a wooden spoon with a spin top on the end and a small hook. It's lightweight, easy to store, and inexpensive, under twenty dollars or so. It is a fun way to learn how to make yarn.

If you're going to process fiber on a large scale, a spinning wheel might be a better option. A lot of people sell used spinning wheels online. You can plan on spending a couple hundred dollars on a spinning wheel.

SELLING

As I mentioned earlier, you can sell raw fiber that has not been washed or carded to those wanting to do the processing themselves or to large textile corporations, or you can do the processing and sell the completed product. Mohair generally sells for $25 for a four-ounce skein; cashgora and cashmere sell for more. Some ideas for where to sell your fiber include specialty shops, online marketplaces such as Etsy, local knitting or crocheting clubs, artisan stores, and more.

ADDITIONAL SERVICES

Here are some other strategies for making money with your four-legged livestock. When deciding what side business you should set up, it's important to think about your market and how big the demand for your products will be. For example, when we lived in the city, we were the only show in town. I could pretty much, within reason, create my own market. I would sell my products and farm supplies at the top of the scale. My eggs and my honey always had a waiting list; it was a matter of supply and demand. Then I moved to the country, where everyone in town raised their own livestock, and I could barely give the stuff away. We moved away from our target demographic. Know where your customers are. There are many places to find customers, including homeschooling groups, health food stores, food co-ops, farmers' markets, restaurants that source food locally, other farmers, online, and more. What is available in your area may inform which of the following ventures best suits you and your herd.

Selling your kids: Goat kids, that is. Resale value is one reason why we decided we wanted a registered herd (see page 14). With papers, we can sell our registered Nigerian Dwarf does for $200 to $700 each. I would get around $150 for an unregistered goat. Bucks tend to sell for less than does, and wethers go for the least amount since they are not used for dairy or breeding purposes. For the past year we have focused on growing our herd to increase our milk supply. Now that we have enough does, all the new babies will be sold this year. If you have four does and each doe kids two offspring, you can make anywhere from $1,600 to over $5,000 in one kidding season. Now you can see the math adding up and the potential to make money from selling kids.

Buck service: If you own a registered buck, you can charge anywhere from $30 to more than $100 per buck service. If your buck comes from show-quality goats, meaning he or his parents placed in goat shows, you can command a higher fee. A one-year-old buck can service ten ladies per season or per month. A two-year-old buck can service twenty-five ladies per month, and a three-year-old buck can service up to forty ladies per season or month! All provided the buck is healthy and in good shape, of course.

Fertilizer or compost: Goat pellets or goat berries are great for the garden. They make wonderful fertilizer that you can sow directly into the soil. Because of its composition, it doesn't burn your plants and you don't have to wait a year to apply it. There is hardly any odor at all, and it doesn't tend to attract flies like cow manure. Gardeners or farmers who work organically are interested in this type of fertilizer. A farmer friend of mine collected goat manure, dried it, bagged it in eight-ounce bags, and sold them for $7.95 each plus tax online and in local nurseries.

Weed control: All over the country in fire-prone areas, people are singing the praises of goats' ability to clear brush to help prevent the spread of forest fires. Depending on where you live, you may be able to rent goats for weed control and clearing brush. A reasonable price would be $200 at base, then

$125 per week to clear one acre, plus food, travel fees, and any other related cost. I've seen others advertise around $1,000 per acre to clear. One important note: because goats are browsers and foragers, not grazers, they do wonders for keeping the brush and weeds at bay, but not so much for mowing the lawn.

Goat therapy: Many animals are used for therapy, helping bring joy, calmness, relaxation, and comfort to humans. Goats are one of those animals that people use to help them feel better. From people who have special needs and seniors to those who suffer with PTSD, goats are offering some much-needed joy. Your goats do not have to be certified to work as therapy animals, they just need to be friendly and enjoy being around people. Another way people are using goats for therapy is by offering goat yoga. All you need are some yoga mats, a yoga instructor, and a bunch of bouncing baby goats. People are charging around $45 for a one-hour session to stretch with goats.

Farm tours and Goat 101: From garden groups to school field trip groups, people of all ages love to visit farms and petting zoos. A goat breeder friend of mine used to offer farm tours and charged $3 per person. She would walk them around the farm, let them pet the goats and observe a milking, and even give some goat milk samples. Once you have some experience under your belt, you can offer classes to other would-be goat owners in goat care, milking, making cheese, making goat soap, and more. Cheesemaking classes are around $50 to $100 per person, and one-on-one goat mentoring would go for around $50+ per hour.

Say cheese: Who doesn't love an adorable spring picture with a baby goat in a field? Spring is the perfect time of year to book photo shoots on your farm with all the new baby goat kids. Photographers and individuals alike would love to book some photo sessions with your baby goats during kidding season. Generally, photo sessions start at $50 per hour for photos with baby goats.

BUDGET BREAKDOWN, EXPECTED REVENUE, AND TIME FRAME

There is no quick or foolproof way to monetize your homestead. Especially when it comes to livestock, you can't rush these things. Whether you're raising goats for dairy, meat, kids, fibers, or all of the above, it will take at least a year of your time before you can start to expect a return on your investment.

Below is an example of your goat expenditures over the course of a year.

Spring

➡ Feed

➡ Fencing

➡ First aid kit

➡ Goat organization membership

➡ Hay

➡ Housing

➡ Purchase of goats

➡ Vet visit for wellness check and vaccinations

Summer

➡ Feed

➡ Hay

➡ Minerals

Fall

➡ Feed

➡ Hay

➡ Minerals

➡ Buck service if you don't have a buck

Winter

➡ Feed

➡ Hay

➡ Minerals

Spring

➡ Feed

➡ Hay

➡ Minerals

➡ Wellness check by vet

➡ Registration for new goat babies

INCOME POTENTIAL

One to two years after your goats have been on your homestead, you can start to monetize them. You may also be able to make some return on your investment before your one-year goat-iversary, depending on the age of the goats you purchased. Here is an example of what that might look like.

Spring Year 2

➤ If you purchased adult fiber goats, you can shear them in the spring ($300+ per goat)

Summer Year 2

➤ Buck service if you have bucks old enough ($40+ per service)

Fall Year 2

➤ Buck service if you have bucks old enough ($40+ per service)

➤ Shear fiber goats ($300+ per goat)

➤ Meat goats are coming to maturity for butcher (6 to 9 months of age for harvest) (market price per pound)

Winter Year 2

➤ Harvest meat goats if you didn't harvest in the fall (market price per pound)

Spring Year 3

➤ Baby goats ($100+ per kid)

➤ Goat milk ($9+ per gallon)

➤ Shear fiber goats ($300+ per goat)

UNDERSTANDING YOUR MARKET

It's important to understand your market. I mentioned earlier that when we lived in the city, there was a huge demand for my products and farm supplies, and they commanded high prices. Then when we moved to the country, we moved away from our target demographic.

Know where your customers are. If you live in the country and want to sell goat meat, you will want to go where people eat or are cooking goat meat.

Some places where you can find customers include:

- Homeschooling groups
- Health food stores
- Nutritionist and doctor offices
- Food co-ops
- Farmers' markets
- Restaurants that source food locally
- Local food hubs
- Farm-to-table movements
- Farm sales
- Crunchy-mama groups
- Facebook
- Craigslist
- Classifieds
- Feed stores—many have boards for advertising
- Local private catering companies
- Goat organizations—many have membership directories with target customers
- Other farmers

appendix

GOAT MILK AND MEAT RECIPES

**W**hether you are raising goats to become more sustainable or you want to make some extra money on the side, your dairy and meat goats will help you provide delicious food for your family and others. These recipes are easy to make, popular with clients who love goat-based food, can help you reach your sustainable goals, and will put some extra money in your family's pockets.

GOAT BURGERS

Prep time: 10 minutes | **Cook time:** 20 minutes | **Serves:** 4

GLUTEN-FREE, NUT-FREE, SOY-FREE

These goat burgers are amazing paired with Whipped Herb Goat Cheese (see page 130). You can serve them on a bun like you would a regular burger or in a lettuce wrap for a truly gluten-free burger. Moist and delicious, this goat burger will make a believer out of you.

2 pounds chuck or ground goat meat

½ small white onion, finely chopped

3 tablespoons finely chopped fresh mint

1 teaspoon sea salt

1 teaspoon freshly ground black pepper

1 teaspoon onion powder

1 teaspoon garlic powder

1 large egg

1 tablespoon grated Parmesan cheese

1 tablespoon Worcestershire sauce

1 teaspoon red pepper flakes (optional)

1. Place the goat meat, onion, mint, salt, pepper, onion powder, garlic powder, egg, Parmesan, Worcestershire sauce, and red pepper flakes (if using) in the bowl of a food processor. Process until well combined.

2. Measure out four 8-ounce portions of the goat meat mixture and form them into patties.

3. Heat a large cast-iron skillet over medium-high heat. (You can also cook these on a grill.) Working in batches if necessary, add the patties and cook on one side for about 5 minutes. Flip and cook on the other side for about 5 minutes, or until a meat thermometer reads at least 160°F when inserted into the middle of the patties.

BAKED GOAT ZITI

Prep time: 30 minutes | **Cook time:** 30 minutes | **Serves:** 4

NUT-FREE, SOY-FREE

This recipe is perfect as a one-dish meal for the family or gatherings. I like to use this recipe for friends who have never tried goat meat. They can never tell the difference and love the taste.

1 tablespoon
   extra-virgin olive oil

1 pound ground
   goat meat

3 or 4 garlic cloves,
   crushed

1 medium onion,
   finely chopped

1 pound ziti or desired
   pasta, cooked
   according to
   package directions

1 (24-ounce) jar
   spaghetti sauce
   of choice

2 cups shredded
   mozzarella cheese

½ cup grated
   Parmesan cheese

1. Preheat the oven to 350°F.
2. Heat the oil in a large skillet over medium heat until it shimmers.
3. Add the goat meat, garlic, and onion and cook, using a spatula to break up the meat, for 10 minutes, or until the meat is browned and cooked through.
4. Remove the skillet from the heat and drain any excess oil.
5. Place the cooked pasta into a shallow 13-by-9-inch baking dish, then add the meat mixture in a layer. Top with the pasta sauce.
6. Top with the mozzarella, then sprinkle with the Parmesan cheese.
7. Bake, uncovered, for 30 minutes, or until the cheese is melted and the sauce is bubbling.

GOAT MEATLOAF

Prep time: 20 minutes | **Cook time:** 45 minutes | **Serves:** 6 to 8

DAIRY-FREE, GLUTEN-FREE, NUT-FREE, SOY-FREE

Meatloaf is one of those comfort foods that most of us grew up with. We all know that one person who made the best meatloaf when we were growing up. With this recipe, that person can be you!

1 teaspoon sea salt

1 teaspoon freshly ground black pepper

1 teaspoon onion powder

1 teaspoon garlic powder

1 large egg

1 tablespoon Worcestershire sauce

½ cup ketchup, divided

3 pounds ground goat meat

½ small white onion, finely chopped

½ green bell pepper, finely chopped

½ cup gluten-free oats or bread crumbs

3 bacon slices, halved

1. Preheat the oven to 350°F.

2. In a large bowl, whisk together the salt, black pepper, onion powder, garlic powder, egg, Worcestershire sauce, and ¼ cup of ketchup until well blended.

3. Add the meat, onion, bell pepper, and oats or bread crumbs, and use your hands to mix thoroughly.

4. Transfer the meat mixture to a 9-by-5-inch loaf pan.

5. Spread the remaining ¼ cup of ketchup over the meatloaf. Place the bacon on top.

6. Place the loaf pan on a baking sheet, then bake for 35 to 45 minutes, or until a meat thermometer registers 160°F when inserted into the middle of the meatloaf.

GOAT CHILI

Prep time: 15 minutes | **Cook time:** 30 minutes | **Serves:** 8 to 10

GLUTEN-FREE, NUT-FREE, SOY-FREE

This recipe is perfect for a cold day. I love adding the leftovers (if there are any) to a baked potato with some cheese and sour cream for a whole new dish!

1 tablespoon extra-virgin olive oil

2 pounds ground goat meat

1 small onion, chopped

½ green bell pepper, chopped

4 garlic cloves, chopped

2 (15-ounce) cans kidney beans (regular or dark), drained

2 (15-ounce) cans pinto beans, drained

1 tablespoon chili powder

1 teaspoon ground cumin

1 teaspoon sugar

1 teaspoon sea salt

Pinch red pepper flakes

1 teaspoon hot sauce

1 (28-ounce) can crushed red tomatoes

1 (15-ounce) can tomato sauce

1 (4-ounce) can tomato paste

Shredded cheddar cheese and sour cream, for garnish

1. In a large stockpot over medium heat, heat oil until it shimmers.
2. Add the goat meat, onion, bell pepper, and garlic and cook, breaking the meat up with a spatula, for 10 minutes, or until the meat is browned and cooked through.
3. Add the kidney and pinto beans, then add the chili powder, cumin, sugar, salt, red pepper flakes, hot sauce, crushed red tomatoes, tomato sauce, and tomato paste, stirring to combine.
4. Reduce heat to low, cover, and simmer for 30 minutes, stirring occasionally.
5. Remove the chili from the heat. Portion into bowls, garnish with cheddar cheese and sour cream, and serve.

SIMPLE GOAT CHEESE

Prep time: 5 minutes, plus 1½ hours to rest | **Cook time:** 15 minutes |
Makes 1 pound

GLUTEN-FREE, NUT-FREE, SOY-FREE, VEGETARIAN

This is one of the reasons why we raise dairy goats: to make our own cheese.
Cheesemaking is never easier than this recipe, and you can do so many things
with it. Spice it up with peppers or add a jam preserve. Enjoy on crackers or
incorporate into your Baked Goat Ziti (see page 126).

½ **gallon goat milk**

⅓ **cup freshly squeezed
lemon juice**

2 **tablespoons white
vinegar**

½ **teaspoon sea salt
(optional)**

1. Line a colander with two or three layers of
fine cheesecloth.
2. In a heavy-bottom saucepan over medium
heat, heat the goat milk to 170°F to 180°F,
stirring frequently to avoid scorching the milk
and to evenly distribute the heat. This will
take 10 to 12 minutes; do not rush it.
3. Immediately remove the milk from the heat
and add the lemon juice and vinegar, stirring
briefly to combine.
4. Set the milk mixture aside for 30 minutes at
room temperature.
5. Slowly ladle the milk mixture into the lined
colander over the sink.
6. Add the salt (if using) and lightly mix.
7. Gather the ends of the cheesecloth and tie a
string around them. Suspend the cheesecloth
bundle over the sink (for example, you can tie
it to the faucet). Allow to drain for 1 hour.
8. Remove the cheese from the cheesecloth
and shape it on a plate.
9. Eat immediately or refrigerate for up to
1 week before enjoying.

WHIPPED HERB GOAT CHEESE

Prep time: 5 minutes | **Makes** ½ pound

GLUTEN-FREE, NUT-FREE, SOY-FREE, VEGETARIAN

This whipped goat cheese goes wonderfully with Goat Burgers (see page 125). It's also amazing as a sandwich spread or on bagels.

8 ounces Simple Goat Cheese (page 129)

1 garlic clove, minced

1 tablespoon chopped fresh parsley

¼ teaspoon finely chopped fresh thyme

¼ teaspoon sea salt

¼ teaspoon freshly ground black pepper

1. Place the goat cheese, garlic, parsley, thyme, salt, and pepper in the bowl of a food processor or stand mixer. Process until well blended.

2. Serve immediately or refrigerate for up to 1 week before enjoying.

GOAT MAC 'N' CHEESE

Prep time: 25 minutes | **Cook time:** 35 minutes | **Serves:** 4 to 6

NUT-FREE, SOY-FREE

I think we can all agree that mac 'n' cheese is a favorite childhood dish that we still love and enjoy as adults. Goat products bring it to the next level. You will enjoy the smokiness of the bacon, the richness of the cheese, and the satisfying crunch of the bread crumbs.

½ cup goat milk

4 ounces Simple Goat Cheese (page 129)

4 ounces shredded cheddar cheese

1 teaspoon sea salt

½ teaspoon freshly ground black pepper

1 pound cooked macaroni

2 large eggs

¼ pound bacon, cut into 1-inch pieces and cooked

1 cup bread crumbs

1. Preheat the oven to 350°F.
2. In a medium saucepan over low heat, combine the milk, goat cheese, cheddar cheese, salt, and pepper and cook, stirring constantly, until the cheese is completely melted.
3. Add the macaroni, stir, and set aside.
4. In a medium bowl, beat the eggs. Fold the eggs into the mac 'n' cheese mixture, then add the bacon and stir to mix.
5. Pour the mixture into a 13-by-9-inch baking dish. Bake, uncovered, for 20 minutes.
6. Remove from the oven and add the bread crumbs on top. Bake for 10 more minutes, or until the top is golden brown.

GOAT CHEESE AND HERB APPETIZER

Prep time: 5 minutes | **Serves:** 4 to 6

GLUTEN-FREE, NUT-FREE, SOY-FREE, VEGETARIAN

This appetizer is simple and easy to make but does not lack pizzazz or wow factor. Serve this goat cheese log with a variety of crackers, a baguette, or sliced apples.

2 tablespoons
   food-grade
   dried lavender

1 (10- to 12-ounce) goat
   cheese log

1 teaspoon freshly
   grated lemon zest

2 tablespoons honey

1. Place a sheet of wax paper on the counter.
2. Sprinkle the lavender on the paper in a line the same length as the cheese log.
3. Place the cheese log on the lavender and roll the wax paper around it to gently press the lavender into the cheese.
4. Transfer the cheese log to a serving tray.
5. Sprinkle the lemon zest on the log and drizzle the honey over it.
6. Serve immediately or refrigerate for up to 1 week before enjoying.

SPRING GOAT CHEESE SALAD WITH RED WINE VINAIGRETTE

Prep time: 10 minutes | **Serves:** 4 to 6

GLUTEN-FREE, SOY-FREE, VEGETARIAN

Goat cheese salad is probably the most familiar way the average American consumes a goat product. The flavors in this salad—the tang of the goat cheese and the sweetness of the dried cranberries—blend well together.

FOR THE SALAD

1 (16-ounce) package spring salad mix

¼ cup finely chopped red onion

½ cup chopped dried cranberries

1 cucumber, peeled, seeded, and sliced

4 ounces goat cheese crumbles

2 tablespoons sliced roasted almonds

FOR THE RED WINE VINAIGRETTE

1 teaspoon minced garlic

1 teaspoon chopped shallot

1 tablespoon Dijon or spicy mustard

½ cup red wine vinegar

1 tablespoon herbs de Provence

1 tablespoon honey or sugar

2 cups vegetable oil

1. **To make the salad:** Toss the spring mix, red onion, dried cranberries, and cucumber in a large bowl.

2. Sprinkle with the goat cheese and almonds.

3. **To make the red wine vinaigrette:** Place the garlic, shallot, mustard, vinegar, herbs de Provence, honey, and oil in a blender and blend until the oil is thickened.

4. Add the vinaigrette to the salad and toss, or serve the dressing on the side.

5. The dressing can be refrigerated for up to 2 weeks. Shake before using.

RESOURCES

GENERAL RESOURCES AND BREED INFORMATION

American Dairy Goat Association (ADGA) | adga.org

American Goat Federation (AGF) | americangoatfederation.org

American Goat Society (AGS) | americangoatsociety.com

American Goat Society, "Breed Standards" | americangoatsociety.com /breed-standards.php

International Fainting Goat Association, "Breed Standards" | fainting goat.com/breed-standards.php

Livestock Conservancy, "Oberhasli Goat" | livestockconservancy.org /index.php/heritage/internal/oberhasli

Livestock Conservancy, "Quick Reference Guide to Heritage Goats" | livestockconservancy.org/images/uploads/docs/GoatChart2019.pdf

Smithsonian's National Zoo & Conservation Biology Institute, "Goat" | nationalzoo.si.edu/animals/goat

USDA Cooperative Extension, "Goat Breeds Nubian" | goats.extension .org/goat-breeds-nubian

GOAT HEALTH, NUTRITION, AND ANATOMY

ACS Distance Education, "Goat Behaviour" | acsedu.co.uk/Info /Pets/Animal-Behaviour/Goat-Behaviour.aspx

Alabama A&M and Auburn Universities Extension, "Digestive System of Goats" | ssl.acesag.auburn.edu/pubs/docs/U/UNP-0060/UNP-0060 -archive.pdf

Backyard Goats, "Kat's Corner: Answers About What to Feed Goats" | backyardgoats.iamcountryside.com/feed-housing/kats-corner-answers -about-what-to-feed-goats

California Department of Food and Agriculture, "*Brucella melitensis* of Goats and Sheep" | cdfa.ca.gov/ahfss/animal_health/pdfs/B _MelitensisFactSheet.pdf

Goat World, "Behavior" | goatworld.com/articles/behavior/behav ior.shtml

Louisiana State University Ag Center, "Aging Goats by Mouthing" | lsuagcenter.com/NR/rdonlyres/F68F5D60-B54F-4476-8966 -1740C6D2B95E/104131/5aginggoatsbymouthingwithtemplate.pdf

Manitoba Goat Association, "Goats and Their Nutrition" | gov .mb.ca/agriculture/livestock/goat/pubs/goats-and-their-nutrition.pdf

MannaPro, "Breaking Down the Goat Diet" | info.mannapro.com /homestead/breaking-down-the-goat-diet

Merck Veterinary Manual, "Tuberculosis in Sheep and Goats" | merckvetmanual.com/generalized-conditions/tuberculosis -and-other-mycobacterial-infections/tuberculosis-in-sheep-and-goats

North Carolina State Extension, "Lice: What They Are and How to Control Them" | content.ces.ncsu.edu/lice-what-they-are-and-how-to-control-them

Oregon State University, "Internal Parasites in Sheep and Goats" | smallfarms.oregonstate.edu/sites/agscid7/files/em9055.pdf

Purdue University Extension, "Common Diseases and Health Problems in Sheep and Goats" | extension.purdue.edu/extmedia/as/as-595 -commondiseases.pdf

Purdue University Extension, "Hoof Anatomy, Care and Management in Livestock" | extension.purdue.edu/extmedia/id/id-321-w.pdf

Scottish Government Department for Environment, Food and Rural Affairs, "Guidance for Tuberculosis in Goats in England and Scotland" | assets.publishing.service.gov.uk/government/uploads/system/uploads/attachment_data/file/371675/TN182.pdf

University of Georgia College of Veterinary Medicine, "Enterotoxemia in Sheep and Goats" | vet.uga.edu/enterotoxemia-in-sheep-and-goats

University of Maine, "Tips for Detecting Disease or Injury in Sheep and Goats" | extension.umaine.edu/publications/1032e

USDA Cooperative Extension, "Goat Nutrition GI Tract" | goats.exten sion.org/goat-nutrition-gi-tract

USDA Cooperative Extension, "Goat Nutrition Water" | goats.extension .org/goat-nutrition-water

USDA Cooperative Extension, "Why Is It So Important to Regularly Trim the Feet of Sheep and Goats?" | animal-welfare.extension.org/why-is -it-so-important-to-regularly-trim-the-feet-of-sheep-and-goats

USDA Sustainable Agriculture Research and Education Program, "Why and How to Do FAMACHA Scoring" | web.uri.edu/sheepngoat/files /FAMACHA-Scoring_Final2.pdf

CASTRATION AND BREEDING

New York State 4-H, "Meat Goat Project Fact Sheet #10" | cpb-us-e1 .wpmucdn.com/blogs.cornell.edu/dist/c/3808/files/2015/02/Castration -25ppfac.pdf

USDA Cooperative Extension, "Goat Reproduction Puberty and Sexual Maturity" | goats.extension.org/goat-reproduction-puberty-and -sexual-maturity

DAIRY GOATS AND GOAT MILK

ADGA Breed Standards | adga.org/breed-standards

Clara Hedrich, "Dairy Goat Management" | milkproduction.com
/Global/PDFs/Bestmanagementpracticesfordairygoatfarmers.pdf

Food and Agriculture Organization of the United Nations, "Factors
Affecting Goat Milk Production and Quality" | agris.fao.org/agris
-search/search.do?recordID=US201400009231

Iowa State University Extension, "Beginner Dairy Goat Fact Sheet" |
extension.iastate.edu/dairyteam/files/page/files/Dairy%20GOATS%20
Fact%20Sheet%2016.pdf

National Park Service, "How Goat's Milk Is Healthier Than Cow's Milk" |
nps.gov/carl/learn/education/classrooms/upload/ES35-post-why
-goats-milk-healthier-info.pdf

PennState Extension, "Dairy Goat Production" | extension.psu.edu
/dairy-goat-production

FIBER GOATS

Agricultural Marketing Resource Center, "Goats for Fiber" | agmrc
.org/commodities-products/livestock/goats/goats-for-fiber

Horace G. Porter and Bernice M. Hornbeck, "Wool and Other Animal
Fibers" | naldc.nal.usda.gov/download/IND43861829/PDF

University of California Small Farm Program, "Angora Goats: A Small
-Scale Agricultural Alternative" | sfp.ucdavis.edu/pubs/brochures
/ANGORA

University of California Small Farm Program, "Cashmere Goats" | sfp
.ucdavis.edu/pubs/brochures/Cashmeregoats

MEAT GOATS AND GOAT MEAT

American Goat Federation, "Meat Goats" | americangoatfederation.org
/breeds-of-goats-2/meat-goats

Michigan State University Animal Legal and Historical Center, "Table of
State Humane Slaughter Laws" | animallaw.info/article/table-state
-humane-slaughter-laws

New York State 4-H, "Meat Goat Project Fact Sheet #19" | cpb-us-e1
.wpmucdn.com/blogs.cornell.edu/dist/c/3808/files/2015/01/factm
g19a-1hbys2c.pdf

North Carolina State Extension, "Nutritional Feeding Management of
Meat Goats" | content.ces.ncsu.edu/nutritional-feeding-management
-of-meat-goats

Oklahoma State University, "Breeds of Meat Goats" | agecon.okstate.edu
/meatgoat/files/Chapter%202.pdf

United States Department of Agriculture, "National Monthly Grass Fed
Lamb and Goat Report" | ams.usda.gov/mnreports/lsmngflamb
goat.pdf

USDA Cooperative Extension, "Marketing Meat Goats, the Basic System" |
goats.extension.org/marketing-meat-goats-the-basic-system

INDEX

ACKNOWLEDGMENTS

If it weren't for Casey Price's sharing her love and knowledge of goats with others, we would have never begun this journey. A special thanks to Shelley Noisette for providing me with our first two does. Thank you to my brother, Robert Larobardiere, and Romanita Stanich for help with the amazing recipes. Lynda D. Kelly and Sam Crandall, without y'all our livestock would probably be six feet under and not thriving. Thank you to all the members of the Old Paths to New Homesteading & Self-Reliant Living group for selflessly sharing your knowledge with others. Above all, heartfelt thanks and appreciation to my daughter Morgan Bradshaw for all the help and support.

ABOUT THE AUTHOR

Amber Bradshaw loves foraging for food, gardening, and cooking outdoors. She teaches others how to become self-sufficient by making things from scratch that are eco-friendly. When she's not in her gardens, you can find her working online.

Amber is the author of *Beekeeping for Beginners*, a former 4-H leader, a blogger, and a public speaker. She and her family filmed the building of their off-grid home for a TV documentary. Amber is happy to share her knowledge with others through public speaking, private instruction, and online at MyHomesteadLife.com.

Goats are an essential part of her sustainable life on her family's developing farm in the mountains of East Tennessee. Her Nigerian Dwarf goats provide fresh milk and breeding stock for income.

CPSIA information can be obtained
at www.ICGtesting.com
Printed in the USA
JSHW021127060821
17516JS00003B/9